D0413295

Experiencing Hubble: Understanding the Greatest Images of the Universe

David M. Meyer, Ph.D.

THE
GREAT
COURSES

PUBLISHED BY:

THE GREAT COURSES
Corporate Headquarters
4840 Westfields Boulevard, Suite 500
Chantilly, Virginia 20151-2299
Phone: 1-800-832-2412
Fax: 703-378-3819
www.thegreatcourses.com

Copyright © The Teaching Company, 2011

Printed in the United States of America

This book is in copyright. All rights reserved.

Without limiting the rights under copyright reserved above,
no part of this publication may be reproduced, stored in
or introduced into a retrieval system, or transmitted,
in any form, or by any means
(electronic, mechanical, photocopying, recording, or otherwise),
without the prior written permission of
The Teaching Company.

David M. Meyer, Ph.D.

Professor of Physics and Astronomy
Northwestern University

Professor David M. Meyer is Professor of Physics and Astronomy, director of Dearborn Observatory, and co-director of the Center for Interdisciplinary Exploration and Research in Astrophysics at Northwestern University. He received his B.S. in Astrophysics at the University of Wisconsin–Madison after completing a senior honors thesis on ultraviolet interstellar extinction with Professor Blair Savage. Professor Meyer earned his M.A . and Ph.D. in Astronomy at the University of California, Los Angeles, working with Professor Michael Jura on measurements of the cosmic microwave background radiation from observations of interstellar cyanogen. He continued his studies as a Robert R. McCormick Postdoctoral Fellow at the University of Chicago's Enrico Fermi Institute before joining the Northwestern faculty in 1987.

Professor Meyer's research focuses on the application of sensitive spectroscopic techniques to astrophysical problems involving interstellar and extragalactic gas clouds. Using a variety of ground- and space-based telescopes, he studies the optical and ultraviolet spectra of stars and quasars to better understand the composition, structure, and physical conditions of intervening clouds in the Milky Way and other galaxies. Over the course of the past 15 years, much of his research has involved data from the Hubble Space Telescope (HST). During this time, Professor Meyer and his collaborators have been awarded more than $1 million in research funding to carry out 20 HST projects that have resulted in 25 peer-reviewed publications on topics ranging from the abundance of interstellar oxygen to the gaseous character of distant galaxies. He has also served 5 times on the committee that annually selects the most deserving proposals for HST observing time.

During his career at Northwestern, Professor Meyer has specialized in designing and teaching introductory undergraduate courses in astronomy, cosmology, and astrobiology for nonscience majors. A hallmark of his lectures is the use of HST images to bring the latest research into the introductory classroom. His success in such efforts has led to a number of teaching awards. In 2009, Professor Meyer was awarded Northwestern's highest teaching honor, the Charles Deering McCormick Professorship of Teaching Excellence. His previous honors include the Koldyke Outstanding Teaching Professorship (2002), the Weinberg Distinguished Teaching Award (1999), and the Northwestern University Alumni Excellence in Teaching Award (1998). He has also reached out to students young and old beyond the Northwestern campus by delivering popular talks on HST science to alumni groups in settings as unusual as a transatlantic crossing of the *Queen Mary 2* in 2005. ∎

Table of Contents

Table of Contents

Experiencing Hubble:
Understanding the Greatest Images of the Universe

Scope:

The Hubble Space Telescope (HST) has revolutionized our understanding of the universe both near and far. Its stunning images of stars, nebulae, and galaxies have captivated public attention and inspired students of all ages. The sharpness of HST's images owes to the telescope's location in near-Earth orbit, well above the blurring effects of atmospheric turbulence. Indeed, HST can routinely image the universe with a resolving power more than 10 times better than that of the largest ground-based telescopes. Since its launch in 1990, HST has taken more than a half million cosmic images and astronomers have published more than 7,000 scientific papers based on HST data. These papers have reported groundbreaking discoveries on a variety of topics, ranging from the formation of stars, to the collisions of galaxies, to the accelerating expansion of the universe.

In this introductory course, we discuss the scientific stories behind 10 of HST's most spectacular images. These 10 images were chosen on the basis of their visual beauty and scientific impact and to illustrate the breadth of HST astronomy. The lectures are organized to address the images one by one from near to far, beginning with the solar system, then on to stars and nebulae in the Milky Way Galaxy, individual galaxies, systems of galaxies, and finally, the universe at large. In each of these lectures, the HST image is discussed in terms of its broad astrophysical context and the specific implications of its findings. Along the way, these "Hubble stories" provide an inside look at the history and operation of HST as it is used to attack the most important problems in modern astrophysics. A key emphasis throughout the course is how HST's unique imaging capabilities have made its discoveries possible. Specific contrasts are drawn between HST's view of the universe and those of the naked eye and ground-based telescopes.

The course begins with an introductory lecture on light and telescopes. It focuses on the advantages of HST over ground-based telescopes and the

Since its launch, the Hubble has taken over a half million images of planets, stars, nebulae, and galaxies, and astronomers have published over 8,000 scientific papers based on Hubble data.

1993 space shuttle servicing mission that installed corrective optics to overcome the spherical aberration of HST's 2.4-meter-diameter primary mirror. We then voyage to the planet Jupiter, whose deep atmosphere was buffeted by the impact of multiple kilometer-sized fragments of Comet Shoemaker-Levy 9 in 1994. The HST image of the temporary Earth-sized scars left by these impacts on Jupiter serves as a contemporary reminder of the continuing impact threat to Earth posed by comets and asteroids. Our first stop outside the solar system is the Sagittarius Star Cloud—one of the richest star fields in the Milky Way. HST's view of the inner star cloud region reveals a sparkling jewel box of colors that is particularly appropriate for a discussion of the wide variety of stars that populate the galaxy. We travel next to the Eagle Nebula to study an active site of star formation inside a vast interstellar cloud of gas and dust. The now-famous HST image of newborn stars emerging from their dusty cocoons at the heart of this nebula was the first to reveal star birth in such detail. Our visit to the Cat's Eye

Nebula provides an opportunity to witness and discuss the final stages in the death of a solar-type star. The multiple gas shells evident in the penetrating HST view of this planetary nebula are the blown-off outer layers of the dying central star. We then explore the violent deaths of the most massive stars through the HST mosaic of the Crab Nebula—the gaseous remnant of a supernova explosion observed on Earth in the year 1054. Such explosions are the primary galactic source of atomic elements, such as oxygen and iron, that make planetary life possible.

The most spectacular HST image of an individual galaxy is that of the magnificent Sombrero Galaxy—a nearly edge-on spiral at a distance of 29 million light-years. Through this image, we discuss the character of galaxies in the local cosmos and the history of their discovery as "island universes" far beyond the Milky Way. We then turn to the deep HST view of the distant galaxies in the sky field of the foreground spiral NGC 3370 to introduce the evidence for an expanding universe. HST observations of Cepheid variable stars and supernovae in NGC 3370 and other galaxies have recently shown that this expansion is accelerating under the mysterious influence of dark energy. Although the overall universe is expanding, the colorful HST image of the colliding Antennae Galaxies shows that gravity can overcome this expansion in regions where galaxies are clustered. We explore this image in the context of what to expect when the nearby Andromeda Galaxy begins to collide with the Milky Way about 2 billion years from now. Some clusters of galaxies are so massive that they curve the surrounding space to an extent that the images of background galaxies are distorted as their light passes through this space. As the most dramatic example of such a gravitational lens, we investigate the HST observations of the cluster Abell 2218 and its implications regarding the dominance of dark matter in the universe. The last of our top-10 HST images is the Hubble Ultra Deep Field—the deepest optical survey of the universe made to date. We discuss how the evolution of galaxies evident in this image is consistent with the idea that the universe began with a Big Bang 13.7 billion years ago. The course concludes with a lecture on the future of HST and the space telescopes to follow in its footsteps, specifically, the James Webb Space Telescope. The lecture focuses on the capabilities of these telescopes in opening up one of the key research frontiers in modern astrophysics—the detection and characterization of extrasolar planetary systems. ∎

The Rationale for a Space Telescope
Lecture 1

One might view this course, actually, as the science equivalent of an art appreciation course. We're going to discuss Hubble's images in the context of the telescope's capabilities and the underlying astrophysics in a similar way that one might discuss an artist's masterpieces in the context of the artist's style and the times during which they were painted.

In this first lecture, we discuss the key advantages of a space telescope over a ground-based observatory in imaging the cosmos. Understanding these advantages requires a brief introduction to the basics of light, telescopes, and the Earth's atmosphere. We will then discuss the attributes of Hubble as designed and how its early performance in Earth orbit nearly shattered its promise as a revolutionary space telescope.

Almost everything we know about the universe beyond the solar system comes from observations of **electromagnetic radiation**. The optical light we see with our eyes is one form of electromagnetic radiation. Light is made of wavelike particles called **photons**, which exhibit an inverse relationship

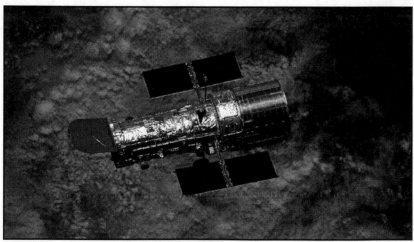

NASA.

Every 97 minutes, Hubble travels around the Earth in a low-Earth orbit.

between their wavelength and energy: Higher-energy photons have shorter wavelengths, and lower-energy photons have longer wavelengths. The shortest-wavelength photons are gamma rays, and from there, the spectrum proceeds to X-rays, ultraviolet light, optical light, infrared light, microwaves, and radio waves.

Each region in the electromagnetic spectrum gives us a different view of the cosmos, and it does this because there's a wide range of processes in the universe that emit radiation at different wavelengths. Our atmosphere is transparent only at optical, radio, and select wavelengths in the infrared and microwave regions of the spectrum. One of the key advantages of a space telescope is that it can observe astronomical objects and processes that emit photons at wavelengths that can't be seen from the Earth's surface.

Most of the telescopes used in astronomical research today are reflectors, which use a curved mirror to focus starlight. The design of a reflecting telescope must ensure that the focus is not in the way of the light path. The **Cassegrain** design, that's used in the Hubble and other telescopes, accomplishes this as follows: A large primary mirror collects starlight, then bounces the starlight off a smaller secondary mirror, which then sends the light through a small hole in the primary mirror to the focus behind the telescope.

Hubble was designed with a complement of instruments to image and take spectra of astronomical objects.

Astronomers are always looking to build telescopes with larger mirrors because a larger mirror collects more light and can resolve smaller angular separations in the sky. From one horizon to another in the sky is 180 degrees. If we used a protractor to measure the angle that the Moon subtends on the sky, we'd come up with 1/2 degree, but astronomers use smaller units of angle: arc minutes (60 arc minutes in 1 degree) and arc seconds (60 arc seconds in 1 arc minute). Using this system, the 1/2-degree Moon is 1,800 arc seconds.

5

If we're looking at two stars separated by an angle of 1 arc second using a telescope with a diameter of 1/10 of a meter, the telescope can just barely resolve those two stars. Our atmosphere, however, poses a problem for increasing resolution with bigger telescopes: The atmosphere is always in motion, and it distorts starlight, causing the effect of twinkling. We might have a huge telescope on the ground, but we're unable to get a sharp image because of this twinkling starlight.

There are two basic approaches to improving **resolving power**. One is to work with mountaintop telescopes using a technology called "adaptive optics"; this technology has had some successes, but there are limits to what it can do, especially beyond tiny fields of view in the sky. The other option is to put the telescope above the atmosphere. The primary rationale for Hubble was the scientific desire for a large telescope in Earth orbit that would be capable of routinely imaging the cosmos with a resolving power more than 10 times better than that of the largest ground-based optical telescopes.

Hubble was put in low-Earth orbit (600 kilometers altitude) for several key reasons, one of which was to enable it to be launched and serviced with the space shuttle. This servicing capability proved to be vital when, shortly after launch in 1990, it was discovered that Hubble's resolving power wasn't much better than that of ground-based telescopes. It turned out that the primary mirror was flawed. But there was also another

The primary rationale for Hubble was the scientific desire for a large telescope in Earth orbit that would be capable of routinely imaging the cosmos with a resolving power more than 10 times better than that of the largest ground-based optical telescopes.

fix needed for an even higher-priority problem. Since launch, astronomers had noted that whenever Hubble crossed the terminator—the point of change from day into night and night into day—the solar panels would flap because they were not rigid enough for night/day heat stress associated with this terminator crossing. The flapping solar panels could break off, causing Hubble to lose power. If Hubble lost power, it couldn't keep the instruments hot enough to function, and the mission would be over.

The 1993 servicing mission made it possible for Hubble to achieve its design goal of imaging the optical universe at an unprecedented sky resolution. It was also a shining success for NASA in demonstrating that astronauts could perform difficult, detailed work in the space environment. ■

Important Terms

Cassegrain telescope: A telescope in which incoming starlight is reflected off a primary mirror to a secondary mirror that then reflects it back through a small central hole in the primary mirror to an eyepiece or instrument behind the primary.

electromagnetic radiation: Commonly referred to as "light" at optical wavelengths, this radiation is due to oscillating electric and magnetic fields.

photon: The particle that carries electromagnetic radiation (light) with wavelike characteristics.

resolving power: A measure of the smallest angular separation that a telescope can resolve in an image.

Suggested Reading

McCray, *Giant Telescopes*.

Smith, *The Space Telescope*.

Zimmerman, *The Universe in a Mirror*.

Questions to Consider

1. What would be the advantages and disadvantages of observing the universe with a large telescope on the Moon's surface as compared to one in low-Earth orbit?

2. Given the success of the HST, why do astronomers continue to build bigger and bigger Earth-based telescopes?

The Rationale for a Space Telescope
Lecture 1—Transcript

Arguably, no scientific instrument in history has had a bigger impact on science and society than the Hubble Space Telescope. Since its launch, the Hubble has taken over a half million images of planets, stars, nebulae, and galaxies and astronomers have published over 8,000 scientific papers based on Hubble data. Groundbreaking Hubble discoveries range from the formation of stars to the collisions of galaxies to the accelerating expansion of the universe. Stunning Hubble images of the cosmos have had appeared everywhere, from postage stamps to popular movies.

Welcome to our course on the Hubble Space Telescope and its view of the universe. My name is Dave Meyer. I'm an observational astronomer and a professor of physics and astronomy at Northwestern University in Evanston, Illinois. I have been preparing, writing, reviewing, and carrying out research proposals for observing time with Hubble ever since the telescope was launched in 1990. It continues to be the most sought-after telescope for observing time among astronomers. Today, we're going to discuss what makes Hubble such a special telescope and how we are going to explore its unparalleled view of the cosmic frontier in this course.

Specifically, we're going to explore the scientific stories behind 10 of what I believe to be the most fascinating Hubble images. These images include a snapshot of the planet Jupiter shortly after it was hammered by a number of kilometer-sized comet fragments in 1994. Each one of these impacts left a temporary dark scar the size of the Earth on the familiar face of Jupiter. Far beyond Jupiter, we will travel with Hubble to one of the many planetary nebulae heralding the deaths of solar-type stars in our Milky Way Galaxy. This image of the Cat's Eye Nebula is a preview of what awaits the Sun in about 6 billion years. We will also venture far into the past with Hubble's Ultra Deep Field image of the most distant galaxies. This image reveals the character of the universe long ago when there was no Earth, there was no Sun, and none of the naked eye stars in the sky had yet been born.

How do you come up with the top 10 images of Hubble? There are so many hundreds, thousands to choose from. I use several criteria in coming up with my top 10. One of the key criteria, of course, was the beauty of the image, the impact; I tried to choose the more stunning images that just jumped out at you as "Wow, this is something special." Science is also very important. I also took that into account in trying to come up with images that either were giving us a view of some exciting new development in astronomy or were giving us a very nice overview, some very basic aspect, of astrophysics. Also, in coming up with these images, I wanted to select 10 that illustrated the breadth of astronomy done with Hubble. I could have easily picked 10 spiral galaxies; I love spiral galaxies, it's easy to pick 10 gorgeous spiral galaxies out of the Hubble gallery. But if I did that, I'd just be shortchanging you because it's just amazing the amount of astronomy that Hubble has done across the spectrum of everything in the universe.

We're going to take kind of a broad view of astronomy with the images that we're going to discuss in this course. In putting these lectures together, I've organized the discussion of the images from near to far. We'll start in the solar system and then move out to stars and nebulae and our Milky Way Galaxy. We'll then talk about individual galaxies, systems of galaxies, and the universe as a whole. The discussion of each image, each of our Hubble stories, will involve the broad astrophysical context of the image; I'll try to put each image, telling you a lot about the astronomy behind it both generally—the general background—and also a little bit more specifically, too, about the astrophysics behind each image. We'll talk about the specific implications of the findings associated with each one of these images; what is it telling us in terms of cutting edge research today?

Along the way, we'll take an inside look at the history of Hubble and operations with the telescope. How do astronomers apply for time with Hubble? How does that work? We will also contrast the Hubble observations with ground-based observations, not only with our largest ground-based telescopes but also with the naked eye. When you look at the sky with your naked eye, what do you see, and how does that contrast with what Hubble can see? One might view this course actually as the science equivalent of an art appreciation course. We're going to discuss Hubble's images in the context of the telescope's capabilities and the underlying astrophysics in a similar way that one might discuss an artist's masterpieces in the context of the artist's style and the times during which they were painted. Throughout this course, a key emphasis will be on how Hubble's unique imaging capabilities have made its discoveries possible.

My own research on interstellar and extragalactic gas clouds has benefited enormously from Hubble observations. I have several ongoing Hubble projects and I have published 25 research papers based on Hubble data since 1994. The topics of these papers range from things like the abundance of interstellar oxygen to the gaseous character of distant galaxies. I've also served five times on the committee that annually selects the top 15 percent of the many proposals for Hubble observing time; and this is really hard work because there are so many excellent observing proposals for Hubble time. All of this work has made Hubble a key part of my research and teaching career over the past 20 years; thus it is a distinct honor and a real pleasure to deliver this introductory course on the science of Hubble.

In this first lecture, we're going to discuss the key advantages of a space telescope over a ground-based observatory in imaging the cosmos. Understanding these advantages requires a brief introduction to the basics of light, telescopes, and the Earth's atmosphere. We will then discuss the attributes of Hubble as designed and how its early performance in Earth orbit nearly shattered its promise as a revolutionary space telescope.

A key thing to recognize right off the top is that almost everything that we know about the universe beyond the solar system comes from observations of electromagnetic radiation. The optical light you see with your eye is one form of this electromagnetic radiation. Light is made of particles called photons; these photons have a wave-like character. You can measure photons; actually, since photons have a wavelength character, you can describe them in physical characteristics such as the physical length between two succeeding crests of a wave. You can measure the wavelength of a photon, and it turns out that higher-energy photons have shorter wavelengths and lower-energy photons have longer wavelengths. There's an inverse relationship between the wavelength of a photon and its energy. Given this, one can describe photons as a function of their wavelength or energy in an electromagnetic spectrum, and actually display photons as a function of their wavelength. For example, let's start with the shortest wavelength photons, the gamma rays; and we can go to succeedingly longer wavelengths and photons from gamma rays to x-rays to ultraviolet to optical light to infrared light to microwaves and radio. Radio photons have the longest wavelengths; the wavelengths of radio photons can be measured in kilometers, they're so long.

When you look at the electromagnetic spectrum, one of the first things you notice is that optical light is just a tiny fraction of the entirety of the electromagnetic spectrum. In terms of wavelength, optical light ranges from 400 nanometers to 700 nanometers, where 1 nanometer is one-billionth of a meter. The wavelengths of optical light are very, very short; and the shorter wave, the 400 nanometer optical light, is violet light, and the 700 nanometer optical light is red light. The colors we see with our eye are essentially we're seeing light of different wavelengths.

The importance of understanding electromagnetic spectrum is that each region in the electromagnetic spectrum gives us a different view of the cosmos, and it does this because there's a wide range of processes in the universe that emit radiation at different wavelengths. For example, the hot, tenuous gas associated with supernova remnants is very bright at x-ray wavelengths, whereas the Sun is very bright at optical wavelengths; and if we look in the cold depths of dense interstellar clouds of gas, we find molecules that can emit radiation at selected microwave and radio wavelengths. This is really important because our atmosphere is only transparent in certain select regions of the electromagnetic spectrum; specifically, only at optical, radio, and select wavelengths in the infrared and the microwave does light actually get through our atmosphere to the ground. What that means is a significant fraction of the electromagnetic spectrum cannot be observed from the ground; and, indeed, until we started putting telescopes up into space in the 1960s, we had no idea what the universe looked like in these other regions, such as gamma rays, and x-rays, and ultraviolet.

Since the 1960s, space astronomy has really blossomed, and astronomers and scientists all over the world have begun to put bigger, better, and more sensitive telescopes in orbit. This whole field of space astronomy has blossomed because now we can study the universe across the electromagnetic spectrum. In my own personal case, a lot of the research I do is in the ultraviolet, and I can't do this from the ground. I need space astronomy to do a good chunk of my research. Thus, one of the key advantages of a space telescope is that it can observe astronomical objects and processes that emit photons at wavelengths that can't be seen from the Earth's surface.

Whether the observations are from space or from the ground, astronomers typically employ telescopes to carry out their work. Like your eye or a camera, a telescope is basically a device that collects light from an object and produces a focused image. Most of the telescopes that are used in astronomical research today are reflectors. These reflecting telescopes use a curved mirror to focus starlight. A good example of such a telescope is the Gemini North eight-meter telescope on the summit of Mauna Kea on the big island of Hawaii (when I say it's an eight-meter telescope, I'm referring to the diameter of its primary mirror). Gemini North—this telescope, the Gemini North eight-meter—and Hubble are similar in that they both have a Cassegrain design to capture the light and bring it to its focus.

The design is important with a reflecting telescope. For example, if you have a small reflecting telescope and the light hits the mirror and comes to a focus, of course, in front of the telescope, if you stick your head over the telescope to look down to see the focus, your head blocks the small mirror; so when you work with reflecting telescopes, you have to get the focus out of the light path. The way the Cassegrain design accomplishes that is you have your big primary mirror that collects starlight, and then the primary mirror bounces the starlight back up and reflects it off a smaller secondary mirror above the primary, and then the smaller secondary mirror sends the light back through a small hole in the primary mirror to a focus behind the telescope. This design is great, particularly if you have heavy equipment—like a heavy spectrograph to measure the spectrum of an object, or a big camera—because if you mount such heavy equipment on the back of a telescope like this, you can balance it more easily and much more effortlessly track objects across the sky as the Earth rotates. At first glance, you would say, "Wait a minute, you punched a hole in your mirror; that can't be good." The point is this hole in the primary mirror leads to only a tiny light loss, and the other gains you get from this design outweigh this tiny light loss.

As I'm sure you're all aware, astronomers are always looking to build telescopes with larger mirrors. Why? Why is this something astronomers want to do? Why do they always want bigger mirrors? There are two points to this: One is because a bigger mirror will collect more light; and also a bigger mirror has a higher resolving power. First, let's think about how a mirror collects more light; and the way to think about this is to think about a telescope's primary mirror as a light bucket. It's collecting photons in the same way a rain bucket might catch raindrops. If you have a bigger rain bucket, a wider rain bucket, you'll catch more raindrops; in a similar way, if you have a telescope primary mirror that is a larger diameter, you'll collect more photons. The photons that you collect will scale with the area of the mirror; such that if I compare a four-meter diameter telescope to an eight-meter diameter telescope, the eight-meter will collect four times the photons because it scales as the area. By collecting more photons in a given amount of time that means you can see fainter stars.

Another key point is that bigger telescopes can also resolve smaller, angular separations on the sky. What do I mean by that? Let's spend a minute talking about angles. You know that in a circle there are 360 degrees; and if we think about that in the context of the sky, from one horizon to another horizon on the sky is 180 degrees. Let's think about angles in terms of the size of certain objects on the sky. In particular, let's look at the moon: The moon, if we got out a protractor and measured the angle that the moon subtends on the sky, we'd come up with an angle of a half a degree. As you can imagine, as we think about smaller angles in astronomy, degree is a kind of an awkward unit; so we can think about tinier units of angle. For example, within 1 degree, there are 60 arc minutes—an arc minute is a tinier angle than a degree—such that the full moon on the sky, which is half a degree, is the same as 30 arc minutes on the sky in terms of an angle. There's an even tinier angle: Within 1 arc minute, there are there are 60 arc seconds; so an arc second is a very tiny angle on the sky, and in units of arc seconds, the half degree moon on the sky is 1,800 arc seconds. Or another way to think about it: An angle of 1 arc second on the sky is $1/1,800^{th}$ the width of the full moon.

Now let's think about two stars: We have two stars in the sky and they're separated by a 1 arc second angle. If you have a telescope with a diameter of a tenth of a meter, it can just barely resolve those two stars; see that there's not one blob there but two blobs. If you looked at those two stars with the naked eye, your eye's iris is much smaller than a tenth of a meter, so what you see is just one fuzzy blob; you don't see two stars with the naked eye, you just see one. But if you had a larger telescope than tenth of a meter, you'd actually more easily split the two stars.

Given this, you might say, "This is great. We'll make even bigger and bigger so we can see the sky from the ground with fantastic accuracy." But there's a problem: our atmosphere. Our atmosphere is a turbulent thing; it's always in motion. Cold fronts, warm fronts, high pressure systems, low pressure systems; the atmosphere's always in motion, and as starlight comes through our atmosphere, it twinkles a little bit (twinkles sometimes a lot). In other words, when you look at the night sky and you see the twinkling stars and you say, "This is beautiful, this is gorgeous; this is wonderful, look at this twinkling of starlight; this is magical," astronomers hate this. The twinkling of stars has nothing to do with the stars themselves; that's our atmosphere distorting the starlight. If we got above the atmosphere, the stars are just points of light, not twinkling at all; it's our atmosphere that makes the stars twinkle, and it's this twinkling that limits a telescope's resolving power. You can have a huge telescope on the ground, but you can't get a sharp image because this twinkling starlight. In other words, this effect limits the resolution of all but the smallest ground-based telescopes.

What can we do about it? A first step is put your telescope on a mountain. That's why the biggest telescopes on the planet are typically on top of mountains. Why? Because if you're on top of a mountain, you're already above a lot of the atmosphere. Indeed, the 14,000-foot Mauna Kea summit on the big island of Hawaii is a fantastic location to put big telescopes, and there are a lot of them up there. So by putting a telescope on top of a mountain you can get above a lot of the atmosphere to help with the seeing, and secondly the airflow across mountaintops is often laminar, not quite as turbulent. But even putting your big telescopes on top of mountains, even as wonderful as the one on Mauna Kea, the seeing there is typically not that much better than one arc second. By "seeing," I mean, "What's the sizable angular disc of a star due to the turbulence of the atmosphere as the starlight comes through?"

There are two basic approaches to improving the resolving power beyond this kind of seeing of a little bit less than an arc second: One, we can work with our mountaintop telescopes and use a certain kind of technology called "adaptive optics" with these telescopes. What this involves is you put special deformable optics in the light path of the light as it goes through the telescope that are under active computer control, and you deform these optics in a way where you can try to partially compensate for the turbulence effect in the atmosphere. This technology is very exciting, but it's also very challenging; it's had some successes, but there are limits to what it can do, especially beyond very, very tiny fields of view in the sky. There's another option: Put your telescope above the atmosphere. Indeed, the idea of a large space telescope dates back to 1946; and the idea is easy conceptually: Hey, just put a telescope above the atmosphere; we don't have to worry about the fuzziness produced by the atmosphere anymore. But space is expensive; indeed, the Hubble Space Telescope in Earth orbit today was shaped by many years of design, debate, and delay over things like science goals, technical issues, politics, and funding.

The primary rationale for Hubble was the scientific desire for a large telescope in Earth orbit that would be capable of routinely imaging the cosmos with a resolving power over 10 times better than that of the largest ground-based optical telescopes. Hubble was designed with a 2.4-meter diameter primary mirror and a complement of instruments to image and take spectra of astronomical objects at optical, ultraviolet, and near-infrared wavelengths.

Let's talk about some of the details associated with Hubble. The first thing you might think about is: Why a 2.4-meter mirror? We just mentioned the Gemini North 8-meter mirror, there are a lot bigger telescopes on the ground; why is the mirror on board Hubble so much smaller? There are several reasons. One is smaller mirrors are cheaper to make; also, they're cheaper to fly because they weigh less, and if they weigh less it doesn't cost as much to get them up into space. Another key point is a 2.4-meter diameter telescope easily fits in the space shuttle, which is important if you want to use the shuttle to deploy it and service the telescope. Another key point is that the United States already has had some experience even before Hubble was put up into orbit building, constructing, and launching 2.4-meter telescopes; they were all used by the military, though, for reconnaissance satellites. Indeed, that's another thing that's worth remembering: In addition to Hubble up there, there are a number of U.S. reconnaissance satellites of the same size;

they're just pointed this way instead of out. But the bottom line reason why a 2.4-meter telescope is because this is what you need to meet the imagine goal. If your goal is to make images of the cosmos a factor of 10 better than our best ground-based telescopes, you don't need a telescope mirror bigger than 2.4 meters.

Another key point about Hubble is—we focused a lot on the telescope so far—what do we do with the photons once our telescope collects them and brings them to a focus? How do we record them? Hubble uses digital cameras to record images. These digital cameras on board Hubble, at their heart, there are detectors—multi-megapixel detectors—that are very similar to the digital cameras you have at home. The difference, though, is that the cameras on board Hubble are kept very cold, and by keeping them cold, and with other key characteristic of these cameras, they can work at very low light levels; in other words, they can detect almost every photon that hits them and work in what is essentially the photon-starved environment of deep astronomy. A key thing about these detectors is they record almost every incident photon; almost every photon that comes in, the detector says, "Got it," and they're recorded electronically so it's very easy to do the quantitative analysis with these electronic detectors that record the data through computers. Such detectors have actually been commonly used with ground-based telescopes since the 1980s.

It's really important to understand that these detectors are much better than the kinds of detectors that astronomers used decades ago, which were photographic plates or photographic film. The efficiency of those detectors was much poorer than these electronic detectors; in other words, these photographic plates only detected one percent of the photons that hit them. Imagine these poor photons that have traveled many, many years to come to Earth, and they actually make it through on a clear night and they actually get through the optics of the telescope but then they go to the photographic plate and whack, we only get 1 out of 100. That's a terrible loss of light. Indeed, this is another key thing to recognize: By using electronic detectors versus photographic plates, we gain a factor of 100 in efficiency in counting photons, which is the same as going to a 10-meter telescope versus a 1-meter telescope.

It's also worth recognizing that the cameras onboard Hubble have a very small field of view, because the telescope's a big telescope. It's a big primary mirror; and the way the telescope is designed, whenever Hubble takes a snapshot it's typically of a very tiny piece of the sky. For example, the advanced camera for surveys currently on board Hubble, whenever it takes a snapshot, it covers an area in the sky equivalent to a little over 1 ½ percent of the sky area of the full moon. The earlier wide field planetary cameras on board Hubble covered only .7 percent of the full moon sky area. Hubble was not designed to survey the entire sky; that's not what it was designed for. What it was designed for was to image small regions of the sky at resolutions of a tenth of an arc second or better.

The last thing I want to comment on regarding Hubble's details is where it flies. Hubble is in low-Earth orbit, 600 kilometers altitude. It was put there for several key reasons: One key reason was so it could be launched and serviced with the space shuttle; so actually the number of times that astronauts have visited the space telescope is, in addition to launching it and deploying it, astronauts have also serviced it five times with the space shuttle. As Hubble goes around the Earth in this low-Earth orbit, it goes around the Earth every 97 minutes. When it's tracking from one object to another object, moving from one part of the sky to another, it moves at the speed as of a clock minute hand; that's as fast as Hubble can point from one star to another. It takes 30 minutes to go from here to here. What this all means—having a big Earth in the sky and it takes this amount of time to move Hubble—is that the efficient scheduling of this telescope is very complicated. That's the major drawback of being in low-Earth orbit; but the key advantage is the ability to service it with the space shuttle should something go wrong.

That was terribly important: This Hubble servicing capability proved to be absolutely vital when shortly after launch in 1990, it was discovered that Hubble's resolving power wasn't much better than one arc second, similar to the best that you could get with the ground-based telescopes; in other words, Hubble didn't improve matters much at all. This was just a stunning disappointment. After many years of careful planning and billions of dollars, astronomers, NASA, the public, everybody; how could this have happened? What went wrong? How can this be; all the money invested, all the effort?

It's easy to forget now given Hubble's many subsequent successes, but back then, Hubble was a national embarrassment; it was a favorite subject for late-night talk show jokes and newspaper cartoons. What went wrong?

It turned out that the primary mirror was flawed. Despite a number of tests and lots of technicians staring at the mirror, this flaw wasn't caught. What was the flaw? It turned out that the outer part of the mirror had been polished down too far. The error was tiny: It was only $1/50^{th}$ the width of a human hair; that's how much it was polished down, just $1/50^{th}$ the width of a human hair. But that was enough to degrade the overall resolution attainable with this mirror to no better than we could get from the ground. The good news—after the initial shock—as astronomers really began to investigate what the problem was it was realized that this "spherical aberration" was a perfect error. In other words, the error produced by grinding down the outer part of the mirror too much is a common sort of error one can obtain with mirrors (spherical aberration), and as a perfect error that means it could be corrected with small corrective lenses for the cameras on board Hubble. Exactly what this error's associated with: What it means is light that hits the outer part of the mirror is brought to a different focus than light that hit the inner part of the mirror; and so if you try to put the focal plane of instruments at any given point, some of the light won't be focused, it will be fuzzy. After it was understood what the problem was, planning began for a 1993 servicing mission with the space shuttle. The idea was: OK, what we're going to do is take the wide field planetary camera on board Hubble and we'll install a new camera just like it but with little corrective optics in it. In addition, the astronauts would put a new instrument called COSTAR on board that would have lots of little lenses that could pop out of it and place in front of the other instruments to correct their vision.

But it turns out there was also another fix needed, and it was actually for an even higher-priority problem. What this had to do with is since launch, astronomers noted that whenever Hubble, as it made its orbit, crossed the terminator—the point where you go from day into night and night into day in the orbit—Hubble would jitter; and this was bad, because it you're taking a long exposure on a distant star or galaxy and the telescope jitters a couple times in orbit that will fuzz out your image. It's another problem with imagine. Eventually, though, trying to understand what this was due to, astronomers realized what was causing this jittering was the solar panels. They were not rigid enough for night/day heat stress associated with this terminator crossing, so they were flapping—these solar panels were flapping—every time they crossed the terminator. This was a serious concern, because with these solar panels flapping, the joints could break off; and if the panels break off, then Hubble loses power. If Hubble loses power, it can't keep the instruments hot enough to function; they freeze up, the telescope's a brick, mission over. So the highest priority on the first servicing mission to Hubble in 1993 wasn't correcting the optics—of course, it was important to correct the optics to restore the original mission, the idea behind Hubble—but a health and safety issue was the most important; they had to replace those solar panels to make sure Hubble had power.

The good news is the servicing mission was a complete success. The new instruments were successfully installed, the solar panels were replaced, and it worked fantastic; and the subsequent observations showed vast improvement. In this image taken with the faint object camera before and after the servicing mission, one can see that before the correction the individual stars that should have been point-like with Hubble had these big fuzzy halos; but after the corrective optics, stars are points. The situation's even more dramatic with galaxy imagines, particularly with Galaxy M100: Galaxy M100 taken with the wide field planetary camera one before the servicing mission, and then with the corrective optics and wide field planetary camera two after the servicing mission. You just note the amazing difference: In the early version, before it was corrected, the galaxy is all kind of washed out. Once you fix the imaging, put in these corrective lenses, now the galaxy's amazingly sharp; you see all kinds of details: the spiral arms, star clusters, dust lines, everything in this galaxy.

Indeed, without touching the flawed primary mirror, the 1993 servicing mission made it possible for Hubble to achieve its design goal of imaging the optical universe at an unprecedented sky resolution of a tenth of an arc second. It was also a shining success for NASA in demonstrating that astronauts could perform difficult, detailed work in the space environment. The spectacular Hubble images that we will examine in this course are a testament to the courage and skill of the astronauts who deployed and serviced Hubble, as well as the efforts of the many scientists and engineers who supported these missions. Our first image will involve the solar system's headline event of the past century. Next time, we'll voyage to Jupiter with Hubble and study the 1994 impact of Comet Shoemaker-Levy 9. Please join us then.

Comet Shoemaker-Levy 9 and Jupiter
Lecture 2

The importance of Shoemaker-Levy 9 is it kind of woke up the astronomical community. It said, "Hey, this is a problem. We just saw Jupiter get whacked; we see that these things have hit the Earth in the past. The rate at which a kilometer-sized asteroid hits the Earth is not zillions of years; it's a couple hundred thousand years. This is something we should get concerned about."

Shortly after the 1993 servicing mission brought Hubble back to where it should have been, astronomers were given an amazing opportunity to witness the solar system event of the century. Over the course of one week in July 1994, 20 fragments of Comet Shoemaker-Levy 9 slammed into

NASA.

In July 1994, 20 fragments of Comet Shoemaker-Levy 9 slammed into the planet Jupiter. Hubble took a detailed image of Jupiter that revealed a series of Earth-sized impact scars across its cloudtops.

the planet Jupiter. As the cometary barrage ended, Hubble took a detailed image of Jupiter that revealed a series of Earth-sized impact scars across its cloudtops, showing us that such an impact could actually happen during our lifetime.

The drama we will discuss today began in March 1993 when Eugene and Carolyn Shoemaker and David Levy discovered their ninth **comet**, which had a "squashed" shape. Immediately after this unusual comet became known to the astronomical community, follow-up ground-based and Hubble observations of the comet were made. These showed at higher resolution that the comet consisted of multiple fragments, and over time, these fragments were slowly increasing in separation. Even Hubble's initial observations of Shoemaker-Levy 9 in 1993, before its vision was corrected, could easily resolve these fragments.

It became apparent that this comet was orbiting Jupiter (which is extremely unusual); that in July 1992, it had actually passed close enough to Jupiter to be broken into pieces by the planet's gravity; and that it would smash into Jupiter in July 1994. With more than a year to plan for this impact, astronomers could marshal the Hubble and other kinds of observatories to witness this particular event. Two months before the impact, Hubble had a new chance to look at Shoemaker-Levy 9 after its vision had been completely restored as a result of the 1993 servicing mission. It produced an amazing image of Shoemaker-Levy 9 showing 21 individual fragments stretching across 700,000 miles of space, and each one of these fragments had its own little tail.

Most Hubble observations require weeks or more of processing and analysis before the "eureka" of discovery can occur, but with the first Shoemaker-Levy 9 impact, a naked-eye glimpse of the raw images told the tale.

Most Hubble observations require weeks or more of processing and analysis before the "eureka" of discovery can occur, but with the first Shoemaker-Levy 9 impact, a naked-eye glimpse of the raw images told the tale. The first fragment of Shoemaker-Levy 9 smashed into Jupiter's night side at a speed of more than 60 times faster than that of a rifle bullet. Even ground-based telescopes showed a fireball rising from behind the planet edge. These observations were consistent with the prediction that these fragments were roughly about a kilometer in size and they would hold together long enough to get below the clouds, explode, then vent the hot gas as a fireball. As the first impact site rotated into sunlight view 90 minutes later, Hubble revealed an impressive, Earth-sized, dark scar on the Jovian cloudtops. Overall, 20 impacts were recorded, with energies ranging from the equivalent of 1 million to 100 million Hiroshima atomic bombs.

At the time of this comet's impact, the astronomical community was uncertain about the number and sizes of **Near-Earth Objects**. A Near-Earth Object is an object that has an orbit that intersects the Earth's orbit; these are the objects that are most likely to hit the Earth. Shoemaker-Levy 9 served as a wake-up call to astronomers. Since its impact, a number of surveys have discovered 6,300 Near-Earth Objects, 800 of which are kilometer-sized or larger. It's believed that this represents about 80 percent of the total number of such large objects. The good news is that these surveys can plot these objects' orbits, and none of the ones identified so far poses a significant immediate threat.

The Shoemaker-Levy 9 impact in 1994 showed us that violent cosmic events could, at least temporarily, dramatically change the face of familiar objects in the solar system during our lifetime. It also demonstrated that Hubble has the power to monitor such events in real time with unparalleled accuracy. ■

Important Terms

comet: A kilometer-sized object of ice and rock that produces a visible tail of vapor and dust as it approaches the Sun during the course of its orbit.

Near-Earth Object (NEO): A nearby asteroid whose orbit intersects that of Earth.

Crovisier and Encrenaz, *Comet Science*.

Levy, *Impact Jupiter*.

Questions to Consider

1. Imagine that the fragments of Comet Shoemaker-Levy 9 had hit Mars instead of Jupiter. What would the impacts and aftermath have looked like as observed with the HST?

2. Imagine that a kilometer-sized asteroid is discovered on an orbital trajectory to impact the Earth in 200 years. What, if anything, could (or should) be done in the next 50 years to try to prevent this catastrophe?

Comet Shoemaker-Levy 9 and Jupiter
Lecture 2—Transcript

Welcome back to our exploration of the universe with the Hubble Space Telescope. Last time, we discussed the key advantages of a space telescope and how Hubble was designed to observe the sky with resolving power over 10 times better than that of the largest ground-based telescopes. Shortly after the 1993 servicing mission brought Hubble back to where it should have been, it took advantage of an amazing opportunity to witness the solar system event of the century. Over the course of one week in July, 1994, 20 fragments of Comet Shoemaker-Levy 9 slammed into the planet Jupiter. There was a great deal of uncertainty beforehand among astronomers as to whether these impacts would provide a visual feast or a fizzle with Hubble. Starting with the first impact, even those expecting a feast were stunned at the breathtaking images.

As the cometary barrage ended, Hubble took a detailed image of Jupiter that revealed a series of Earth-sized impact scars across its cloudtops. It's still real shocking for me today to see this familiar view of Jupiter's red spot and its atmosphere marred by these dark scars. On this basis alone, I think that this image belongs on any top 10 list on Hubble. But even more so, this image helped to raise worldwide awareness that the planetary impact threat posed by comets and asteroids is more than the stuff of science fiction movies and ancient history. Comet Shoemaker-Levy 9 showed us that such an impact could actually happen during our lifetime and Hubble was actually there to capture the drama in real time. Today, we're going to explore the Hubble view of this comet's demise in the broader context of the many such objects in the solar system and their impact threat to Earth.

Along with the Earth and seven other planets, there are a vast number of meteoroids, asteroids, comets, and minor planets—including Pluto—of various sizes, shapes, and constitutions that orbit the Sun. Although the planets are well-separated, this environment is crowded enough with these other objects that collisions are inevitable. Most of these collisions, by far the bulk of them, are completely insignificant; and you've probably actually seen some of them. These insignificant collisions, the best way to see them is in the context of a meteor or a "shooting star"; it appears as a momentary streaking flash across the sky. If you haven't seen this phenomenon, where you should go is go to a dark site on a clear night—what I mean by a dark site is a site far from any city lights; 50 miles away from any city lights—and go there on a night when the moon isn't up but it's very clear. After you go to such a site and get dark-adapted—it takes your eyes about 20 minutes to get dark-adapted—I guarantee you'll see a meteor within 10–15 minutes. What is one of these meteors due to? It's the result of the impact of something the size of a grain of sand—a little rock the size of a grain of sand—smashing into our upper atmosphere; and such a thing is moving very fast: A typical meteor is moving over 10 times faster than a rifle bullet. When this sand-grain-sized little rock hits the upper atmosphere, it burns up and we see this flash. This happens frequently enough on Earth such that literally 10,000 tons of meteoritic material falls to the Earth every year.

Going much larger in scale than a typical meteor, a typical comet has, say, the size of a kilometer, and it has a core of ice and rock. A typical comet is essentially a dirty snowball; and it's big enough that it can make a significant planetary impact. There are lots of comets in the solar system—it's estimated that trillions of comets orbit the Sun—but almost all of them orbit far, far away from the Sun in the cold depths of the solar system beyond the orbit of Neptune. Comets are essentially leftover material from the formation of the solar system. For the most part, they stay out there in the outer solar system and come nowhere near Earth and Mars and the other planets in the inner part of the solar system; but occasionally way out there, a comet can be gravitationally perturbed by a passing distant star or by a minor planet or something else and a comet will come falling into the inner solar system. As the comet approaches the Sun—remember, this is like a dirty snowball—its outer ices will vaporize, and around this typically kilometer nucleus of rock and ice, as this ice begins to vaporize it will start creating a cloud of gas and dust around this nucleus that can be much, much bigger than a nucleus. A "coma" is what we call this cloud around the nucleus of a comet, and this coma can reach a size of 100,000 kilometers in diameter.

As this comet gets closer and closer into the Sun, the Sun continues to work on the comet. There's a wind of particles coming from the Sun that we call the "solar wind." As the solar wind plus the radiation from the Sun works on the comet, this solar wind will blow the comet and create the comet's characteristic tail. That's one of the things that ever since childhood you know a comet has a tail; the tail is produced due to the interaction of the solar wind with the comet. The key thing to recognize is the tail of a comet always points away from the Sun, it is not indicative of motion; so, indeed, if I have a comet with a tail here that doesn't mean the comet can go in this direction, in the direction of its tail. The tail can be quite spectacular: The tail can be as long as the distance between the Earth and the Sun, 93 million miles.

We get these comets quite frequently in the inner solar system, but many of them are not that bright, certainly to be seen with the naked eye. But every five years or so, we get a comet that's bright enough to be seen with the naked eye, and they can be quite spectacular. The most spectacular comet I've ever seen in my life was Comet Hyakutake in 1996. I happened to be observing at Kitt Peak National Observatory about 60 miles southwest of Tucson, and I had an observing run at the four meter telescope in March, 1996. When you're observing one of these big, ground-based telescopes you're inside this big dome, actually in a control room. You're not looking at the sky at all; you're busy monitoring instruments, you're watching a TV-view of a tiny little piece of the sky that the big telescope's pointing at, but you really aren't paying much attention to the sky as a whole, the whole night sky. I really wasn't there to observe the comet, I was doing some other science entirely; but I was taking long exposures with my instrument so I could take a break. I took a break in the middle of the night and I went out, and I was just stunned at Comet Hyakutake; literally its tail stretched across half the sky. It's just an amazing sight, one of these comets, and indeed it should be no surprise to anyone who has seen one of these bright comets in all its spectacular glory on a truly dark night that throughout human history these comets have really been taken as a herald of doom or great joy of all kinds from different peoples seeing this kind of thing in the sky.

What happens to these comets? First, a bright comet can be a great sight in the sky for up to weeks or a month at a time. It's very different than a meteor. A meteor is a flash across the sky, whereas a comet can be up for this length of time as it slowly moves with respect to the stars. As comets come into the inner solar system, they can have a range of fates. Some will pass so close to the Sun they'll literally fall right into the Sun. Others can be thrown right out of the solar system, never to return. Most of them return way back out to the outer depths of the solar system on very long orbits; they may not return for hundreds of thousands or millions of years back to the inner solar system. But some, a few, are perturbed by the planet Jupiter into orbits in the inner solar system. Perhaps the most famous example of such a comet is Halley's Comet, which comes around every 76 years and has been doing so for a couple thousand years.

Why is Jupiter so important when it comes to comets? Jupiter can seriously perturb the orbits of nearby comets because it's the largest planet in the solar system; Jupiter has more mass than all of the other planets combined. This particular Hubble image of Jupiter shows it partially eclipsing its largest moon, Ganymede, which is about half the diameter of Earth. This image should give you a size of the respective scale of Earth and Jupiter. Jupiter's a very different planet than Earth. Not only is it much more massive—300 times more massive—than the Earth, but Jupiter is a gas giant; it's a big gas ball. Thus, you can't land on Jupiter; if anyone invites you on a trip in a spacecraft to go land on Jupiter don't go, because what basically happens is if you tried to land on Jupiter, you and your ship would just sink deeper and deeper into the atmosphere, and the atmospheric pressure of Jupiter increases with depth and eventually you'd get to depth where your ship and you would be crushed. As you go even deeper inside Jupiter, you eventually see a dense liquid and we expect at the very core of Jupiter there is a solid there. But the point is Jupiter is this big gas ball, and it's a big planet.

When we look Jupiter, we look at the surface clouds of Jupiter; and from what we can see with the eye and the telescopes, they're about 50 kilometers deep. We see different colors that are a function of the cloud altitude at a particular latitude, and a composition of the clouds at these different altitudes. We see these alternating bands across the face of Jupiter, and they're marked by jet streams (gas moving at high velocity); indeed, these winds alternate direction and can move at speeds over 200 miles per hour. One of the most remarkable things about the surface of Jupiter that we can see with the eye is this Great Red Spot. It's a giant atmospheric hurricane. It's about twice the size of the Earth and it's persisted on Jupiter for at least 150 years.

Jupiter plays a big role in the drama we're going to discuss today because there's a story behind the way Jupiter and a particular comet came to meet each other; and we're going to talk about this discovery in March, 1993 by Eugene and Carolyn Shoemaker and David Levy. They discovered an unusual "squashed" object during a comet/asteroid hunt with a small telescope at Palomar Observatory in California. These comet hunts, what they're all about is you take a telescope that looks at kind of a large part of the sky and you compare what the view of one part of the sky is at one time, and then you come back hours or days later and take another picture of the same part of the sky. The relative positions of the stars don't change as time goes by, but any comet, anything in the solar system indeed, would appear to move with respect to those stars over the course of hours to days or weeks. The Shoemakers and David Levy found this object, and it was just so weird: It didn't look like a comet with a tail; it just looked like this squashed-looking thing. It turned out this became Comet Shoemaker-Levy 9 because this was the ninth comet discovered by this team, and it turned out to be even more special than its appearance.

In order to really understand a comet, you have to figure out what its orbit is; and one observation will not give you an orbit. Immediately after this unusual comet became known to the astronomical community, there was follow-up ground-based and Hubble observations of the comet; and it showed at higher resolution that it consisted of multiple fragments and over time these fragments were separating even more over time, slowly increasing in separation. Even though Hubble's initial observations of Shoemaker-Levy 9 were in 1993 before its vision was corrected, even in that state Hubble could easily resolve these fragments. As more and more observations were carried out, the orbit of this comet began to become clear; and it was an astonishing set of results.

First of all, the key result right off the bat was that this comet was orbiting Jupiter. It wasn't orbiting the Sun; it was just orbiting Jupiter in a very elliptical orbit. That's extremely unusual; almost all the comets in the inner part of the solar system orbit the Sun, and then they go out beyond Mars or Jupiter or Saturn, but here was a comet just orbiting Jupiter. Then it became clear that this comet had actually passed very, very close to Jupiter in July, 1992; so close that Jupiter's enormous gravity could take that comet and break it up into pieces, and that accounted for this squashed-looking effect that the Shoemakers and David Levy found. But best of all, what the orbit told astronomers was in predicting the future of this comet, it was clear as early as May, 1993 that this comment was going to smash into Jupiter in July, 1994.

The good news about this: By having over a year to plan for this impact, astronomers could begin to marshal all the observatories on the planet and the Hubble and other kinds of observatories to think about observing this particular event. The problem, though, right off the bat was all the impacts would be occurring on Jupiter's night side; so it would not be possible to directly observe these comet impacts from Earth. However, they would be impacting close enough to the night-day edge of Jupiter so that any fireballs rising above the planet could perhaps be seen from Earth. Furthermore, Jupiter rotates rapidly—like once every 10 hours—so whatever happened after it hit the cloud tops of Jupiter that would rather quickly revolve into the view of our telescopes and Hubble at Earth. Two months before the impact— and indeed observations kept going on of this comet right up until July, 1994—Hubble had a new chance to look at Shoemaker-Levy 9, and this was at a point in time where its vision had been completely restored as a result of the 1993 servicing mission. It produced this amazing image of Shoemaker-Levy 9 showing 21 individual fragments stretching across 700,000 miles of space, and each one of these fragments had its own little tail.

A comparison between this Hubble image and the earlier Hubble image and other images showed that as time went on, this comet was continuing to break up. We already had some information that these fragments were somewhat loosely held together, which we kind of expected anyways because that would explain how it broke up in the first place when it came close to Jupiter, because it was kind of held loosely together. But there was still a lot of uncertainty, even with these wonderful images from Hubble, what was going to happen when this comet smashed into Jupiter. There were two key uncertainties: One, how big were the fragments? Not even Hubble could resolve the fragment sizes. The largest nuclei among these fragments could easily be kilometers in size or perhaps much less; and the reason even Hubble couldn't tell us is because each one of these nuclei was hidden by a large coma of dust and ice.

So we didn't know how big they were—we knew they were roughly a kilometer, not exactly how big—and also the constitution of the fragments, how tightly they were packed together, was clearly unknown. This was totally vital to what kind of event you would get when these things smacked into Jupiter. For example, if each fragment was put together like a rubble pile—just like lots of little pieces stuck together—as it got close to Jupiter just before impact, it might break apart into lots of little pieces; and if lots of little pieces hit Jupiter, you're not going to see any big fireball. This is the way it could effectively be a "fizzle," and this was certainly possible given the Hubble image of an early breakup. On the other hand, if it was a much more solid kilometer in size, it could tunnel right through the atmosphere and go deep enough to a point where it would heat up, vaporize, and just explode and then blow hot gas back through this tunnel up into the upper atmosphere and beyond and it could be seen as a very huge, visible fireball. This is the kind of thing that many astronomers wanted to see: a big explosion. On the third hand, if this thing was a very solid little kilometer, it could tunnel so deep into the atmosphere by the time it blew up it might not come back up through the atmosphere and we'd see any signature.

There was a lot of uncertainty about what we were going to see. Given this uncertainty, astronomers at the Space Telescope Science Institute excitedly gathered around a computer monitor to see the first Hubble images of the impact of this first Shoemaker-Levy 9 fragment. This is very unusual. Most Hubble observations require weeks or more of processing and analysis to perhaps arrive at a "eureka" moment of new discovery. The first Shoemaker-Levy 9 impact was that rare case where just a naked-eye glimpse of the first raw images could tell the tale. I love this image, because what it shows is scientists enjoying the moment of discovery. This is what science is all about: Waiting for this event, you have really no idea what's going to happen, and just looking at this raw image you're going to know what happened; just looking at this one image. In the course of a scientist's career, these kinds of events are very rare; this is the idea of predicting a phenomenon, making guesses, and then actually seeing what happens. I've had this happen to me at telescopes completely serendipitously. You'll be observing some object, getting the data, and you look at the data and it's completely not what you expected; and you've realized you've discovered something new, and it's just amazing. The rush with this is almost indescribable; it's comparable for a scientist to like hitting a home run to win the World Series or throwing a touchdown pass to win the Super Bowl. This is the kind of thing that scientists live for, and that's why in this image you see so many people so excited about this result.

They weren't disappointed, because as the first fragment of Shoemaker-Levy 9 smashed into Jupiter, it smashed into Jupiter at a speed over 60 times faster than a rifle bullet. Both Hubble and ground-based telescopes did indeed show this fireball rising from behind the planet edge. These observations were consistent with the prediction that these fragments were roughly about a kilometer in size and they would hold all together long enough to get below the clouds, explode, and then vent the hot gas as a fireball above. As the first impact site rotated into view 90 minutes later, Hubble revealed an impressive Earth-sized dark scar on the Jovian cloudtops. This is really impressive. Think about this: This rock maybe a kilometer in size whacks Jupiter, and you see this scar shortly thereafter the size of the Earth on the surface of the planet Jupiter. Later analysis showed this dark material consisted of various dust and ices. It has a brownish color, probably rich in sulfur and nitrogen compounds. Where did this stuff come from? Where did the dark stuff come from?

It came mostly from the comet, we think—this thing goes down, explodes, and the comet material comes rushing up to the top of the atmosphere—but also it may well have dredged up some of this atmospheric material in Jupiter's deeper inner atmosphere; and furthermore, just the interaction between this comet ejecta and the material from the inner part of Jupiter coming up may have actually interacted with each other and produced new compounds. As we look at each one of the scars left as a result of the individual impacts of the fragments, we note that they all have roughly similar geometries just after the impact: They all have a dark central core, an inner ring, and an outer crescent that's in the direction from which the thing hit. We think that these characteristics of each impact scar is a part of both the ejecta rising up from the hot fireball cooling off and then settling back on the clouds, and also the impact itself into the upper atmosphere of Jupiter kicked off atmospheric waves that led to interactions between the Jupiter material and the comet material.

Near the end of the barrage of all these fragments on Jupiter, Hubble snapped our feature image showing the dark scars at 8 of the impact sites. Overall, there were 20 impacts recorded, each one having impact energies ranging from the equivalent of a million to 100 million Hiroshima atomic bombs. If you look at this image, you see that the scar shapes are somewhat different from one image to another. All these impacts happened in the course of a week, just one week; and what you're seeing here, these different scar shapes, is the rapid evolution of these scars due to the temporal effects associated with the impact itself and these high-speed winds in the Jovian atmosphere. Remember, there are winds of 200-plus miles per hour and sometimes they're moving in opposite directions in nearby bands, and they can shear these scar material off so they quickly degrade back into the basic colors and banded structure of Jupiter. If we look close up at one of these dark scars, this one's particularly interesting because in this same location after this one fragment hit, a few days later there was another fragment; so here you're seeing the co-evolution of two impact scars over the course of a few days to a week and up to a month, and you can see here this rapid diffusion of the material in with the basic Jovian atmospheric material. Indeed, after about a year there was little trace left of any of these impacts on the upper atmosphere of Jupiter.

When we see these images that Hubble took of the result of this impact of the comet on Jupiter, what are the lessons? What are the lessons of Shoemaker-Levy 9 for Earth? On the one hand, Jupiter is a much bigger target than Earth for comet and asteroid impacts. You think: Well he's a big guy, he's going to get hit more frequently, so this isn't all that unexpected; and he can take it, after a year or so goes by there's not anything left really from what happened. On the other hand, something like Shoemaker-Levy 9, if it had hit the Earth, it would have much more significant longer-term effects on Earth than on Jupiter. Yes, Earth's a much tinier target, and it's much less likely to get hit than Jupiter; but Earth is a big rock, and it could have much more significant effects on Earth unlike Jupiter. Indeed, on Earth, such an impact would not only wreak unimaginable devastation in localized areas, but could also lead to serious global climate change and a mass extinction of many life forms. This has happened before on the Earth; indeed, if we look back 65 million years ago, a 10 kilometer-sized asteroid smashed into Earth and led to the extinction of the dinosaurs (indeed, 60 percent of the life forms on the planet). What happened, there certainly was a tremendous loss of life associated with the impact itself, but the key factor was this asteroid kicked up so much ash and dust into the upper atmosphere that it blanketed the Earth and limited the amount of light that made it to the surface, and as a result all kinds of life forms died; it trickled right up the food chain. We know this has happened in the past, and we know it will happen in the future; the question is when?

At the time of the Comet Shoemaker-Levy 9 impact, there was great uncertainty about the number and sizes of Near-Earth Objects. A Near-Earth Object is an object that has an orbit that intersects the Earth's orbit; these are the objects that are most likely to hit the Earth. I talked about asteroids; and a continuum of comets t asteroids where comets are ice and rock, asteroids are essentially rocky objects, and they have sizes ranging from meters to hundreds of kilometers. The asteroids are typically found in belts, like the asteroid belt between Mars and Jupiter, and also there's another belt of asteroids beyond Neptune in the Kuiper Belt. These asteroids in the asteroid belt between Mars and Jupiter through interactions can occasionally end up on orbits that intersect the Earth's orbit. The question is that Shoemaker-Levy 9 stimulated: What are the odds that one of these things could hit the Earth?

We certainly can know that in the vicinity of the Earth that there have been a number of impacts in the past besides the one caused the extinction of the dinosaurs; you just have to look at the Moon. When you look at the Moon, you see that its craters testify to many past impacts. But how does that compare to the Earth? If we look at the Earth, there are many fewer craters on the surface of the Earth? Why is that? Is there something special about the Earth compared to the Moon? It turns out the Earth has an atmosphere, a very significant atmosphere, unlike the moon, and the Earth has very significant weathering effects and an active geology. In most places where a crater occurs as a result of an impact, the erosion due to weathering occurs and a crater doesn't typically last all that long. Understand it's the cratering, by looking at these craters and comparing their sizes and ages, one can get an estimate and idea of the frequency of these hits. The most well-preserved meteor crater on Earth is the Barringer Crater near Winslow, Arizona. This crater has a diameter of 1 kilometer. It was produced by just a 30-meter-wide asteroid 50,000 years ago. How could just a 30 meter rock produce this big of a hole? Because it's moving so fast; it has a tremendous amount of energy. Indeed, the impact energy of this rock was equivalent to 200 Hiroshima atomic bombs. By studying this crater and other craters, and other evidence like evidence for air bursts of smaller objects in Earth's upper atmosphere, one can estimate the impact rate. As one might expect, smaller objects are more likely to hit the Earth than larger objects; larger objects are rare. But the larger objects, of course, are the ones we're most concerned about because they can cause significant damage. Specifically, how about kilometer-sized Near-Earth Objects? Based on the crater analysis, we see that these objects typically hit Earth every few 100,000 years; and the concern about a kilometer-sized Near-Earth Object is that it can cause continent-wide destruction.

The importance of Shoemaker-Levy 9 is it kind of woke up the astronomical community; it said, "Hey, this is a problem. We just saw Jupiter get whacked; we see that these things have hit the Earth in the past; the rate at which a kilometer-sized asteroid hits the Earth is not zillions of years, it's a couple hundred thousand years. This is something we should get concerned about." But prior to Shoemaker-Levy 9, there were only a few people all over the planet working the problem, equivalent to maybe the staffing of a small fast food restaurant. Since Shoemaker-Levy 9, as a result of this impact, there has been a number of surveys that have gone out and discovered literally 6,300 Near-Earth Objects; and among these 6,300, 800 of them are kilometer-sized and greater. It's believed that this represents about 80 percent of the total number of such large objects. The good news is that in addition to identifying these asteroids, these surveys can also plot their orbits and figure out whether or not they pose a threat to Earth; and indeed the good news is none of the ones identified so far certainly pose a significant immediate threat, certainly not in the next 100 years and most likely far beyond that.

You might say, "Well, so let's say we found one that would hit us in 200 years, what can we do about it?" If we catch one that early, we could theoretically send a mission of some kind and use some kind of technology to just tweak that orbit a little bit of that asteroid; if we just tweak the orbit a little bit now, it would miss the Earth in 200, because Earth is still a tiny target to hit. Indeed, it might be possible for humans to avoid the fate that the dinosaurs suffered; we're smarter than the dinosaurs, we can stop one of these things. As if we needed any further motivation to get concerned about this, Jupiter sent us evidence of another impact scar in July, 2009. A backyard astronomer in Australia found a new dark scar with his equipment—his own personal telescope, his own personal detectors; he just saw this dark spot— he was observing Jupiter, one of his favorite objects, and he found this. One of the most important things to note in astronomy is there are on the order of 20,000 professional astronomers worldwide (astronomers with a Ph.D.), but there are far more astronomers who have an avid interest in astronomy and scan the sky with their favorite objects and make real contributions. When this amateur astronomer found this object, astronomers using Hubble were told about it, they took a picture, and sure enough it's an impact scar much like Shoemaker-Levy 9 16 years ago, and the impacting object probably was hundreds of meters in size. But unlike Shoemaker-Levy 9, no one saw this incoming object beforehand.

The latest event on Jupiter reinforces the importance of the Hubble image of the multiple Shoemaker-Levy 9 impacts in 1994. In summary, this image showed us that violent cosmic events could, at least temporarily, dramatically change the face of familiar objects in the solar system during our lifetime. It also demonstrated that Hubble has this power to monitor such events in real time with unparalleled accuracy. Most importantly, it has helped to stimulate action toward a better understanding of the impact threat to Earth.

The rest of our course is going to focus on Hubble images of the cosmos far beyond Jupiter and the solar system. Next time, we're going to begin our tour of the Milky Way Galaxy with a visit to the Sagittarius Star Cloud. Please join us then.

The Sagittarius Star Cloud
Lecture 3

The total sky area encompassed by this image [of the Sagittarius Star Cloud] is half a percent of the full Moon sky area; just a tiny fraction of the Moon's sky area is in this image, effectively—12,000 stars. That's four times more stars than if you went outside at night and counted all the stars you could see with your eyes.

In this lecture, we leave the solar system and begin to explore the Milky Way Galaxy with Hubble. Our first stop is the Sagittarius Star Cloud, one of the richest star fields in the night sky and located near the center, or bulge, of our galaxy. The Hubble view of the star cloud looks like a sparkling jewel box in the sky and illustrates one of the major advantages of Hubble over ground-based telescopes: its ability to image the cosmos at very high spatial resolution and to distinguish closely spaced stars. With a ground-based telescope, the light from many of the stars in this dense star cloud will blend together. That's a problem if we want to try to compare the properties of nearby stars to more distant stars. The Hubble is able to clearly show stars in the Sagittarius Star Cloud in a wide variety of colors and brightnesses.

The brightness of a star is a function not only of how bright it truly is (its luminosity) but also its distance. If we know the luminosity and brightness of a star, we can calculate its distance. We can estimate a star's intrinsic luminosity by using the **Hertzsprung-Russell (HR) diagram**, which describes stars in terms of their surface temperatures and luminosities. Using the HR diagram, we can easily measure the color of a star, then infer its temperature and luminosity. Combining that information with a measure of the star's brightness, we can get the distance of the star.

By resolving out the colors and brightnesses of individual stars in the Sagittarius Star Cloud and other dense star fields, Hubble facilitates the application of the HR diagram to determine the distances of stars and estimate their ages. Older stars are red and cooler; younger stars are blue and hotter. The Hubble image reveals that the distant stars in the galactic bulge have a redder color distribution than those stars near the Sun, indicating that the stellar population near the bulge consists of older stars than those in the solar neighborhood.

The ages of stellar populations in the galactic halo can also be probed through studies of its **globular star clusters**. There are 150 of these globulars scattered throughout the galactic halo—that is, outside the spiral and center of the galaxy. The dense star fields of these globulars are an attractive target for the Hubble. The results can be surprising: The HR diagram for the cluster M80, for example, reveals that this cluster is more than 10 billion years old, yet Hubble's superior resolution has discovered blue stars at the core of this old globular, a surprise that has astronomers proposing that old stars collide and combine to make hotter ones.

By resolving out the colors and brightnesses of individual stars in dense fields, like the Sagittarius Star Cloud and globular clusters,

The HR diagram for the cluster M80, for example, reveals that this cluster is more than 10 billion years old, yet Hubble's superior resolution has discovered blue stars at the core of this old globular, a surprise that has astronomers proposing that old stars collide and combine to make hotter ones.

Hubble is helping to chart the stellar history and evolution of the Milky Way. Clearly, this evolution is keyed by the nature, timing, and extent of star formation in various regions of the galaxy. ∎

Important Terms

globular star cluster: A densely packed, spherical cluster of up to a million old stars typically found in the halos of galaxies.

Hertzsprung-Russell (HR) diagram: A diagram comparing the temperatures and luminosities of stars that is useful in charting their evolution over time.

Suggested Reading

Kaler, *Extreme Stars*.

————, *The Hundred Greatest Stars*.

Sparke and Gallagher, *Galaxies in the Universe*.

Questions to Consider

1. How would the night sky look to the naked eye on Earth if there were no dust clouds in the Milky Way Galaxy? How would the night sky look if the Earth was located at the center of a globular cluster?

2. How would the HST view of a dustier region of the Milky Way disk contrast with that of the Sagittarius Star Cloud? How would the HR diagrams of these two regions compare?

The Sagittarius Star Cloud
Lecture 3—Transcript

Welcome back to our discussion of the Hubble Space Telescope and its view of the universe. Last time, we voyaged to Jupiter and revisited the 1994 impact of Comet Shoemaker-Levy 9 with the planet. It was truly the solar system event of the past century. Today, we leave the solar system and begin to explore our Milky Way Galaxy with Hubble.

Our first stop in this adventure is the Sagittarius Star Cloud, one of the richest star fields in the night sky. I chose the Hubble view of the Star Cloud as a top 10 image for several reasons: First and foremost, it's simply beautiful; it looks like a sparkling jewel box in the sky. Second, it illustrates one of the major advantages of Hubble over ground-based telescopes: the ability to image the cosmos at very high spatial resolution and separate out closely-spaced stars in dense starfields. Finally, it clearly shows that stars come in a variety of colors and brightnesses; although frankly this image does not represent one of Hubble's greatest discoveries in terms of its science, it does serve as a beautiful introduction to the structure of our galaxy and the varied properties of its stars.

The best place to start when discussing the Galaxy is a naked eye view of the sky on a clear, moonless night from a truly dark site; which you'll notice right away, as compared to a view from like in a city, you see many, many more stars. But another thing that will become obvious rather quickly as you get dark-adapted is that the stars are not scattered randomly across the sky. Indeed, what you'll notice is that there's a band of light stretching across the sky. That band of light is what we call the Milky Way. When you look at the Milky Way, you're actually looking at an edge-on view of the disk of our galaxy. It's the home to most of the 300 billion stars in the galaxy.

The most spectacular view I ever had of the Milky Way was back in the mid-1980s when I was observing at Cerro Tololo Inter-American Observatory in Chile on the west coast of South America. This observatory is located at an altitude about 7,000 feet; it's in the foothills of the Andes Mountains. The view from CTIO is fantastic: About 35 miles to the west is the Pacific Ocean, and rising to the east are the unbelievable peaks of the Andes Mountains rising up to 15,000 feet and higher. This particular night, I was observing at the four-meter telescope, and it was a spectacular night; it was absolutely clear. The seeing was fantastic; the images were extraordinarily sharp. There was no moon in the sky, and it was just completely clear. I was doing some long exposures with the spectrograph on the telescope, so I decided to take a break and go outside and see what the sky looked like. I get out and look at the sky, and the first thing I notice when I get outside is that the small cities you usually see from the observatory along the coast of Chile were dark; there had been a power failure, and all the cities in the vicinity of Cerro Tololo along the coast were completely dark. The only lights I could see anywhere were in the sky, and the Milky Way itself was just blazing across the sky. I could almost imagine, as I stared long enough at the snow-capped peaks of the Andes, in their direction, I could see a glimpse of their snow-capped peak right off the starlight alone. But what really struck me, though, that time, just staring at the Milky Way, was I could realize that I was part of the Milky Way; that along with the Earth and the Sun here, we're part of the galaxy along with the other 300 billion stars in the Milky Way. Indeed, the thought that occurred to me: If there was someone else on another planet around another star far, far away in the disk of the Milky Way, that person at the very same time might be getting a very same view in their night sky of this spectacular Milky Way stretching across the sky.

When observing the night sky, one of the most important things to understand is that all of the stars in the Milky Way are much, much further away from Earth than the Earth is from the Sun or from any of the planets in the solar system. Let's spend some time putting these vast distance differences in perspective. First of all, the distance between the Sun and the Earth is 93 million miles. That's a big number; it's an astronomical number. Sometimes big numbers like this are so big they lose their meaning. Another way to think about it is: How long does it take to travel between the Earth and the Sun? Let's pick a rather low speed, like the speed that your car would take. Let's say you could get your car up to 100 miles per hour and you decided to drive to the Sun if that were possible. How long would it take? It turns out that's 100 years; it takes 100 years to drive to the sun at 100 miles per hour. What's the fastest thing in the universe? It's light; nothing travels faster through space than light. Light moves through space at a speed of 300,000 kilometers per second. And even at that fantastic speed, it takes eight minutes for light to travel from the Sun to the Earth.

The amount of time it takes light to travel a certain distances is the basis for the way astronomers consider distances in the universe. In this kind of system, the distance between the Sun and the Earth would be said to be eight light-minutes. This is a very important unit; there's a very important idea behind this unit. What that means is: When I look at the Sun, I don't see it as it is now, I see it was it was eight minutes ago. It's not possible for me to know what's going on at the Sun right now. Even if I put a satellite in orbit around the Sun, and if something happened at the Sun please send me a radio message to let me know that something had happened on the Sun, radio photons are just like optical photons, they travel at the speed of light; the message would take eight minutes to get to Earth. There's no way to know. The fundamental idea behind these units is that the universe we see is the universe of the past, and how far back we're looking into the past is a function of how far away we're looking out into the universe.

In these kinds of light travel time distance units, the planets in the solar system are all light-minutes to light-hours away. The stars, though, are much, much further away. The nearest star to the Sun is Alpha Centauri, and as the nearest star, it's 4.3 light-years away. In other words, there's this enormous gulf of space between the stars much, much, much more vast than that between the planets and the solar system. A few words about Alpha Centauri; we'll be talking about names of stars and other objects in the universe throughout this course, and I just thought I'd say a few words about Alpha Centauri itself: It's the brightest star in the constellation Centauri. Ancient peoples grouped the stars they observed in the night sky according to their traditions, their mythology, and they'd tell stories based on what they saw in the sky. They would group certain regions of the sky—it would be of this constellation, and that constellation—and astronomers today still use these star groupings to refer to stars in different places on the sky; it's like a navigational aide. Also, a very basic way to categorize the brightest stars in a constellation is use the Greek lettering system; such that Alpha Centauri is the brightest star in the constellation Centaurus, Beta Centauri is the second-brightest star in constellation Centaurus, etc.

When we look at these stars and we see that they're so far away—when you stare out at the sky at night and you look at those stars—it's really important to realize the importance of these distances. They're so far away there's only one way we can visit them, and that's by telescope. We can certainly—and we have—build spacecraft that can visit the other planets in our solar system, light-minutes to light-hours away; but light-years away, no mode of propulsion that we can think of, and we'll be able to think of any time for the far-foreseeable future, can get us to travel light-year distances in a human lifetime. Therefore, the only way we're going to learn about the stars is from the light that we can gather with our telescopes. I also want to emphasize again that the universe we're observing is the universe of the past. A light-minute or two difference may not make that much difference between planets in the solar system, but when we look at Alpha Centauri, the nearest star, we're seeing it as it was over four years ago. Conversely, if someone at Alpha Centauri was there and they were looking at Earth right now, they wouldn't be seeing the Earth of today, they'd be seeing the Earth as it was in the past, four years ago. Indeed, as you go further and further away in the galaxy, the view right now of Earth is even further into the past. This is a fundamental idea central to all of astronomy.

Since we live inside the Milky Way Galaxy and it is so vast, we can't get a "bird's-eye" perspective on its size and structure from the outside. Nevertheless, through a variety of techniques including observations of other galaxies, we've learned a lot about the Milky Way's structure over the past century. First of all, we know it's a spiral galaxy; in other words, if we viewed the galaxy face-on we'd see it has spiral arms. But the galaxy's actually quite thin, relatively speaking: The diameter of the Milky Way viewed face-on is about 100,000 light-years, but viewed edge-on, the stellar disk of the Milky Way is only about 1,000 light-years thick. The Sun is not at the center of the Milky Way Galaxy, it turns out it's far away in the suburbs; the Sun is in the galactic disk, about 28,000 light-years from the galactic center. In addition to the disk of the Milky Way, there are other basic parts of its structure. Around the center of the Milky Way, there's a bulge of stars; it extends out to about 3,000 light-years. The Milky Way has a halo encompassing the disk of the Milky Way; and in the halo, it's very sparsely populated by stars, except there are certain regions, certain "star islands" (we call "globular clusters"), that can be very rich in stars. We'll talk more about those at the end of today's lecture.

One of the things that is quite evident when we look at the Milky Way in the sky: We see dark patches; we see absences of stars. Is it the result that there are no stars there, or is something else going on? It turns out the absence of stars is due to dust clouds in the line of sight. These dust clouds restrict on average our optical view of the disk of the Milky Way to about 2,000 light-years on average. Note, however, that if you're looking outside the disk, above and below the disk, into the halo, you're not restricted so much by dust; indeed, you can see well into the halo, indeed much further out into extragalactic space. As a result of this dust in the disk of the Milky Way, it's really hard to make deep views into the disk of the galaxy; but there are a few dust-poor regions, and very importantly the Sagittarius Star Cloud, the topic of today's lecture, lies in such a region. It's a bright patch of light near the center of the Galactic Bulge. Importantly, it provides a view right into the inner part of the galaxy. A key motivation to peer so deeply into the Sagittarius Star Cloud window is to compare the properties of stars in our neighborhood around the Sun with those in the central Galactic Bulge region of the Milky Way.

This Hubble image sampling the Sagittarius Star Cloud is simply beautiful; there are so many stars in this image, and they're all so colorful. What you're seeing here is you're looking at stars at a variety of distances out to the Galactic Bulge at distances beyond 25,000 light-years. The primary reason that there are so many stars in this image is that we are seeing stars not only in the middle range but all the way out past the Galactic Bulge. In a dustier region of the Milky Way where the optical view might be limited to just a few thousand light-years, a similar Hubble exposure would reveal many fewer stars, and thus you wouldn't be seeing quite as far away in such a field as well.

Only Hubble can adequately resolve out the individual stars in this dense Star Cloud field for study. At lower resolution, like from a ground-based telescope, with such a rich star field, all these stars, a lot of them will blend together; and so what you have effectively is some of these nearby and some of the more distant ones will blend together, make several blend into one.

That's a problem: If you want to try to compare the properties of nearby stars to more distant stars, if their light blends together that makes it much more difficult. Given this, it's really instructive to put the star density of this Star Cloud image in perspective. If you took the time to count every star in this image, you'd come up with 12,000—there are 12,000 stars in this image—but the sky area that's covered is tiny: The total sky area encompassed by this image is half a percent of the full Moon sky area; just a tiny fraction of the Moon's sky area is in this image effectively. 12,000 stars: That's four times more stars than if you went outside at night and counted all the stars you could see with your eye. This is an amazingly dense star field as taken with Hubble. Note, however, that not one of the stars in this Hubble image is naked-eye detectable; all these stars are fainter than the faintest star you can see with your eye. One of the reasons this image is so impressive; how can Hubble see so many stars? It's not just because it has a big mirror, much bigger than the iris of our eye, but also because it's a long-time exposure photograph; your eye basically takes an instantaneous snapshot of what it looks at. But here, if you have a camera that's capable of integrating all the photons that hit your detector for minutes to hours at a time you can build up and detect all kinds of faint things in the universe.

This Hubble Star Cloud image clearly demonstrates that stars come in a wide variety of colors and brightnesses. This image is a composite of three sets of individual observations taken through red, green, and blue filters. Each set involves a total exposure of about 35–50 minutes. The key thing about a filter is that it only lets through light of a specific color. As a result, if you use a filter to take an observation, you specifically can tell among the objects that you're observing—let's say you use the blue filter—which is bright in the blue, which object's faint in the blue, etc.; and then if you take observations through a red filter, you do the same thing with red: what's bright in red; what's faint in red. Then you can combine all these images together—the red, the green, and the blue images—into a composite; and by doing so, by putting together a composite of these primary colors, you can effectively simulate any color viewable by your eye. This works in same way as a TV camera does.

What does this image tell us about the stars? First, the color: The color of a star reflects its surface temperature. Hot stars appear blue because they emit more blue light than red light; cool stars appear red because they emit more red light than blue light. The brightness of a star is a little bit more complicated because, for example, if you see a faint star, what are you seeing? Are you actually seeing a truly faint star that's close by, or are you seeing actually a really bright star that's very far away? In other words, what you're seeing here is the brightness of a star is a function not only of how bright it truly is (its luminosity), but also its distance. Also, dust can come into play, too, if you're looking in a sightline that has a lot of dust in it; that can make a star appear fainter. But let's just focus on distance and luminosity. First of all, as you kind of know from your own experience, the brightness of an object drops as the distance increases; the amount that it drops is proportional to the distance squared. For example, if I double the distance to a star, its brightness goes down by a factor of four; if I triple the distance to a star, its brightness goes down by a factor of nine. If I know the luminosity and I measure the brightness—and measuring brightness is oftentimes much easier than guessing what the luminosity is—if I know those two quantities, I can calculate the distance of a star. But how do we estimate a star's intrinsic luminosity? That's the real puzzle.

What we do is we utilize something called the Hertzsprung-Russell or HR diagram. This diagram describes the characteristics of stars as a function of their surface temperatures and their luminosities. Since these characteristics change as a star ages, the HR diagram can also be used to chart the life histories of stars. When stars are formed in dense clouds of dust and gas, they're made mostly of hydrogen—indeed, most of the stuff in the universe is made out of hydrogen—and what happens shortly after a star begins to form, deep in its core, the core gets hot enough and at high enough pressure where it can start hydrogen nuclei fusing into helium nuclei through a series of nuclear reactions that produce energy that powers the stars. It turns out that stars spend most of their lives deriving their energy from the nuclear fusion of hydrogen into helium at their cores. During this time, there is a relationship between the surface temperatures and the luminosities on the HR diagram denoted by a band called the main sequence. If we look at the HR diagram, we see as plotted here the temperatures are listed in degrees Kelvin; degrees Kelvin is equal to degrees Celsius plus 273 degrees. On the Kelvin temperature scale, water freezes at 273K, and our own body temperature of humans, we have a body temperature of 98.6F, but in the Kelvin temperature scale it's 310K.

If we look at the main sequence on the HR diagram, what we notice: The hottest main sequence stars, which are the bluest main sequence stars, are the most luminous; the hottest main sequence stars are the most luminous. They also turn out to be the most massive. You might think: Well, they're the most luminous, they're the most massive; they must lead really long lives. No; they lead the shortest lives of all: The most luminous, most massive main sequence stars lead lives on the order of a million years; they live fast and die young. On the other hand, the coolest main sequence stars, which are the reddest, are the least luminous and they also tend to be the least massive; and, concurrently with noting that the highest-temperature stars lead the shortest lives, these coolest stars lead the longest lives: They can lead lives of 100 billion years, greater than that, on the main sequence. Where does the Sun fit in this continuum of stars? Its surface temperature is about 5,800K. It's a mid-main sequence star, and it appears white in space with a roughly equal mix of colors. The sun will have a main sequence life expectancy of about 10 billion years. What you can do with all these stars on this HR diagram is you can use it to get the luminosity of a star from a star's color. The star's color is something you can measure: You can measure the color of a star; and from that color you infer its temperature, and if it's on the main sequence you can use that location on the main sequence to infer its luminosity. With that luminosity, then, if you can measure the star's brightness—which is relatively easy to do—you can get the distance of the star. So the HR diagram can be extraordinarily useful in not only getting the luminosity of a star but also its distance.

The groups of stars above and below the main sequence in the HR diagram constitute later stages in the lives of main sequence stars. The stars above the main sequence are giants and supergiants, and those below the sequence are called white dwarfs. A red giant is much more luminous than a red main sequence star with the same surface temperature because it's much larger; a bigger star is more luminous of the same temperature, that's why they're called "giants" or "supergiants." Likewise, a white dwarf is much less luminous than a main sequence star with the same temperature because it's much smaller. These HR diagrams, taking this information into account, it turns out you can use the evolutionary data embedded in an HR diagram to estimate the age of a star group.

What do I mean by a "star group?" A lot of the stars we see in the Milky Way Galaxy are in clusters; and these clusters can range from just a few stars, tens, to hundreds, to literally hundreds of thousands, to millions. The key thing about a star cluster is the stars in the star cluster are believed to have been born at essentially the same time; and so if you have a group of stars that were all born at the same time, you can utilize the HR diagram of this cluster to learn something about its age. For example, let's say we looked at a star cluster and we determined its HR diagram, and we found that every star in this cluster was somewhere on the main sequence, running from the bluest stars to the reddest stars, and we saw no white dwarfs, or no giants, or supergiants. We would know that this is a very, very young star cluster. Why? Because the blue stars, which would lead very short lives on the main sequence—only about a million years—are still there; they haven't had time to evolve off, so this has to be a young star cluster. Let's say we looked at another star cluster, and this other star cluster has no blue stars at all; there's lots of other stars—if we go down the main sequence, we see other stars—but the blue stars are absent, they just aren't there. The way to understand that: Let's say also we looked at that and we saw some supergiants. That would tell us that this cluster is old enough, it's more than a few million years old, enough time for the blue stars to have evolved off the main sequence and become supergiant stars. At the very opposite end of this scale, let's say we look at a cluster and we see no blue stars, no yellow stars, indeed all we see on the main sequence are these red stars, and we see lots of red giants and some white dwarfs. We know this would be a very old star cluster, maybe 10 billion years or more old.

By resolving out the colors and brightnesses of individual stars in the Sagittarius Star Cloud, and other dense star fields, Hubble makes it possible to better apply the HR diagram in determining the distances of the stars and estimating the ages of these stellar groupings. Specifically, if we analyze the HST Star Cloud image, what we find is the distant Galactic Bulge stars have a redder color distribution than those stars near the Sun. The lack of these younger, bluer stars in the Bulge region tells us that near the Bulge it consists of an older stellar population than does the stars in the solar neighborhood.

The wider range of colors and ages among the stars nearer the Sun is consistent with our observations of ongoing star formation in the dense interstellar clouds in the galactic disk. Except for regions very close to the Galactic Center, there is less star formation going on in the Galactic Bulge than in the galactic disk; thus, there are very few young blue stars observed in the Star Cloud view of the Bulge. Near the Sun, however, we have a lot of interstellar matter in the disk of the galaxy at the Sun is from the galactic center, and it's this interstellar matter that keeps the star formation going. As stars die, they return a lot of the material back into the interstellar medium, and these clouds of gas and dust eventually form new stars, essentially recycling the ashes of dead stars into new ones; a cosmic recycling program. The Sun itself is a middle-aged star; by middle-aged, it's about 4.6 billion years old. It's really important to understand this: The Sun, the Earth, and us people are basically recycled stardust; everything we're made out of has passed through earlier generations of stars that formed and died before the Sun ever formed.

The ages of stellar populations in the galactic halo can also be probed through studies of its globular star clusters. A typical globular contains about hundreds of thousands to perhaps millions of stars in a spherical cluster about 100 light-years across. There are 150 of these globulars scattered throughout the galactic halo, typically at distances greater than 10,000 light-years. The dense star fields of these globulars are an attractive target for the Hubble Space Telescope. A really fascinating globular to look at is the globular Messier 80, for short called M80, and it has a particularly dense core. In this Hubble image of this globular M80, there are literally hundreds of thousands of stars. Let's take a minute to contrast that with the view of the Sagittarius Star Cloud with Hubble: When we look at the Sagittarius Star Cloud, we're also looking out very far—out to distances of like 28,000 light-years—but we're seeing stars in the Sagittarius Star Cloud at various distances along the way; some nearby stars some 10,000 light-years away, and some way out to the Galactic Bulge beyond 25,000 light-years. They're just distributed throughout this distance space. But in the case of this globular cluster M80, essentially all the stars we're seeing toward M80 are in M80; remember, the halo is very sparsely populated by stars. So here, essentially every star we see in this image is associated with this globular cluster.

When we make an HR diagram for all the stars in this image, we're essentially making the HR diagram for this cluster M80; and when we do such an analysis, we find that this cluster has an age over 10 billion years. But the ironic thing is Hubble, with its superior resolution, can peer deep inside the core of this dense cluster and separate out all these stars—separate out the glow, and separate these stars—and it's actually found blue stars at the core of this globular. How does this make sense? How could there be blue stars in a cluster that's over 10 billion years old? From everything we know about stars and what we know about the HR diagram, the blue stars that were in this cluster originally should have evolved away off the main sequence long, long ago; many billions of years ago. How can there be blue stars in the core of this cluster? Astronomers call such blue stars "blue stragglers," and the way we understand what's happening is the core of a cluster like M80 is so dense that stars will run into each other occasionally; there will be collisions between stars, they'll spiral into one another. If you have these old stars that spiral into each other—old, cool, red stars—when they become one star, this new star is more massive, hotter, and thereby blue. These two stars have essentially found the fountain of youth, and the reason they exist is because of the high density deep inside the core of this particular cluster, and the same phenomena is seen at the cores of other globular clusters.

Another fantastic globular cluster observed by Hubble is Omega Centauri. Omega Centauri, if you look at it from the ground with your eye or a small telescope, looks just like a single star. But when you look at it with a big telescope, particularly with Hubble, you see that this single bright star becomes many stars; indeed, this globular cluster has five times more stars than M80, almost a million stars. Furthermore, the density at the core of this cluster is amazing: it's 10,000 times higher than that near the Sun. Imagine the view of the night sky on a planet around a star at the core of Omega Centauri. You would have literally tens of thousands, maybe even 100,000, stars in the sky, all of them bright; the nighttime sky would just be glowing with brightness, a truly amazing sight. The HR diagram of Omega Centauri indicates that this cluster also has an age greater than 10 billion years; indeed, most of the globulars in the Milky Way have these kinds of ages What that's telling us is globulars are among the Galaxy's oldest objects. All of their stars formed in a quick burst over 10 billion years ago. As a result of that burst, it left little interstellar matter in the cluster for later star formation.

These globular clusters are ancient relics of an earlier time in the history of the Milky Way.

It's really remarkable that we can learn so much from the colors and brightnesses of stars. The Hubble image of the Sagittarius Star Cloud can both enchant us by its beauty and teach us something about the stars in different parts of the Galaxy. By resolving out the colors and the brightnesses of individual stars in dense fields like the Sagittarius Star Cloud and globular clusters, Hubble is helping to chart the stellar history and evolution of the Milky Way. Clearly, this evolution is keyed by the nature, the timing, and the extent of star formation in various regions of the galaxy. Next time, we will discuss the most famous of Hubble's spectacular images: a breathtaking view of star formation in progress at the heart of the Eagle Nebula. Please join us then.

The Star Factory Inside the Eagle Nebula
Lecture 4

> How were [the Eagle Nebula] pillars sculpted? Nearby hot stars, as their radiation and stellar winds were working on the inside of this molecular cloud, evaporated the less dense material around the pillars, slowly eating it away and uncovering the denser parts, uncovering these dense stalks, uncovering these globules, and beginning to reveal ... that there were stars forming inside.

Some of the most breathtaking Hubble images have involved its observations of interstellar clouds at unprecedented resolution. Among these Hubble cloud images, none has been received with greater public acclaim than the detailed 1995 snapshot of newborn stars emerging from giant pillars of gas and dust inside the Eagle Nebula.

Any wide-field view of the Milky Way across a dark sky reveals the existence of obscuring dust clouds in the disk of the galaxy. These clouds of dust and gas contain in total about 15 percent of the galaxy's visible mass, with the gas making up the bulk of this fraction. The grains of stardust absorb and scatter background starlight; for this reason, dense clouds of gas are also typically dark clouds. In these individual clouds, even though the densities are low, atoms can run into each other and form

NASA, ESA, M. Robberto (Space Telescope Science Institute/ESA), and the Hubble Space Telescope Orion Treasury Project Team.

The Orion Nebula can be seen as a faint whitish glow with the naked eye. The illumination of the Orion Nebula comes almost entirely from four really hot stars called the Trapezium Stars deep inside the nebula.

molecules. Astronomers typically refer to dark clouds as molecular clouds because they have great numbers of molecules deep in their cores. We find molecules ranging from formaldehyde to ethyl alcohol; indeed, some molecules have as many as 13 atoms in them.

These dense regions of dark molecular clouds are important because they contract under their own gravity, their interiors slowly heat up until core hydrogen fusion begins, and a star is born. Active star formation is taking place in the Orion Nebula, the most visible of these gas/dust clouds, but most of this formation is obscured by the very dusty cloud. We depend on infrared and radio telescopes to tell us what's going on in there.

Active star formation is taking place in the Orion Nebula, the most visible of these gas/dust clouds, but most of this formation is obscured by the very dusty cloud. We depend on infrared and radio telescopes to tell us what's going on in there.

Fortunately, the Eagle Nebula provides an optical view inside a large molecular cloud that has produced thousands of stars. The wide-field nebular glow of the Eagle Nebula covers a sky area greater than the full Moon. This glow is excited by the ultraviolet light from an interior cluster of 50 hot young stars that were born inside the cloud. The winds and the radiation from these stars have shaped this cloud over millions of years; they've helped to open a 20-**light-year**–wide window into the cloud so that we can see its sculpted interior with Hubble and other ground-based telescopes at optimal wavelengths.

In 1995, Hubble took our close-up image of the prominent gas/dust pillars at the core of the Eagle Nebula and revealed the wealth of illuminating small-scale structure. The image reveals details on the tallest pillar—thin protuberances—some of which appear to have stars emerging from the globules at the tips. To give a sense of scale, the globules here are 400 times the Earth-Sun distance, and the tallest pillar is 4 light-years in length. The globules are essentially cocoons of gas and dust for the formation of stars.

Stars form as a result of the contraction of dense cloud regions. As a cloud rotates, it contracts (the principle of conservation of angular momentum), and as it gets smaller, it also forms an accretion disk around the developing inner star. The disk funnels matter into the central protostar. Sometimes this matter can come in too fast for the protostar to swallow; in this case, the matter is channeled out into bipolar outflows from the forming star. The disk that's left over from star formation may eventually lead to the formation of planets. The bottom line is: Star formation leads not only to stars but also to jets, disks, and perhaps, planet growth. ■

Important Term

light-year: The distance (6 trillion miles) that light travels in one year.

Suggested Reading

Kaler, *Cosmic Clouds.*

O'Dell, *The Orion Nebula.*

Questions to Consider

1. How might the wide-field and HST views of the Eagle Nebula lead us to conclude that the star formation process in a large molecular cloud is rather inefficient in converting interstellar gas and dust into stars?

2. Why has the HST image of the Eagle Nebula become so popular?

The Star Factory Inside the Eagle Nebula
Lecture 4—Transcript

Welcome back to our exploration of the universe with the Hubble Space Telescope. Last time, we examined the Sagittarius Star Cloud and discussed the wide variety of stars and their distribution in our Milky Way Galaxy. It turns out the space between these stars is not empty; it contains interstellar clouds of various sizes, shapes, densities, and temperatures. These clouds are sculpted by a variety of physical processes ranging from gravity to stellar winds to stellar radiation. Some of the most breathtaking Hubble images have involved its observations of such cloud sculptures at unprecedented sky resolution.

Among these Hubble cloud images, none has been received with greater public acclaim than the detailed 1995 snapshot of newborn stars emerging from giant pillars of gas and dust inside the Eagle Nebula. This image looks like a fantasy landscape out of a dream or a children's story, but it's really in the sky for anyone to see with a telescope like Hubble. I remember how flabbergasted I was when I first saw this image; it was the first time that I fully realized the degree to which Hubble could show us a universe beyond our imagination. By providing a detailed view of star formation in action, the science behind this Hubble image is just as revealing as its beauty. In today's lecture, we're going to explore this image in the context of galactic interstellar clouds, the process of star formation, and the impact of young stars on their interstellar environments.

As we discussed last time, any wide-field view of the Milky Way across a dark sky reveals the existence of obscuring dust clouds in the disk of the galaxy. It turns out that these clouds of dust and gas contain in total about 15 percent of the galaxy's visible mass, with the gas making up the bulk of this fraction. If we count up the composition of the gas—in other words, in terms of the elements in the periodic table that make it up—we find that 90 percent of the gas is hydrogen, about another 9 percent is helium, and 1 percent is all the rest of the elements in the periodic table. In addition to this gas, we have dust grains in the inner stellar medium. Most of these grains are tiny, they're like submicron-sized, and these grains are made out of compounds of things like carbon, oxygen, silicone, magnesium, and iron. I'd like to spend a minute or two talking about how we know this about the dust and the gas since that involves a lot of my research with the Hubble Space Telescope; and indeed, in this course we've been focusing a lot—and we will focus— on images done with Hubble, but there are also great instruments on board Hubble that do spectroscopy.

The instrument I use for my research is the Space Telescope Imaging Spectrograph. A lot of the work that I do is study the ultraviolet spectra of stars, not so much for the stars themselves, but about what I can look at in these spectra to see the signatures of various interstellar atoms in clouds between us and these stars. Each element has a fingerprint in these spectra, so by studying these ultraviolet spectra stars and looking for interstellar absorption features associated with different atoms in the gas, I can measure the abundances of the different atoms in interstellar clouds; and then by making assumptions about what the overall composition of the interstellar medium is, roughly like the Sun, I can infer how much is in the dust. When we add up in total all these cold, dust, and gas clouds, whether it be tenuous or rather dense, together these cold clouds of dust and gas fill up about two percent of the volume of the Milky Way Galaxy. The rest of the interstellar medium is a warmer, tenuous, intercloud medium. The overall density of gas in the interstellar medium is about one atom per cubic centimeter; that's amazingly empty, that's much less dense than the best lab vacuums on Earth. The best lab vacuums on Earth typically have densities on the order of 10,000 atoms per cubic centimeter. There are places in the interstellar medium that are much denser than the average; specifically, we find the cores of some dense, cold clouds to have gas densities reading a million atoms per cubic centimeters. But even at a million atoms per cubic centimeter, the densest interstellar cloud is still much, much, much less dense than the Earth's atmosphere; specifically, the density of Earth's atmosphere at sea level is 13 orders of magnitude greater than the density of the densest interstellar cloud.

These interstellar clouds—those clouds that are very gas-rich—are typically also dust-rich. Gas goes along with dust; so where you have a lot of gas, you typically have a lot of dust. They're roughly pretty well-mixed. (I should interject here as well: I keep referring to dust, and your kneejerk reaction is to think, "Dust, that's something we sweep up, we find under the bed; dust bunnies, etc." That's not as elegant as I'd like it to be. Think of dust as stardust, it's a much more elegant concept. So these dust grains, indeed, they're essentially produced as a result of stars, and we'll more about that later in the course. But whenever I say "dust," I mean "stardust.") These dust grains of stardust absorb and scatter background starlight; therefore, when you have a dense cloud of gas, it's also typically a dark cloud, because it also is a lot of dust and these dust grains block the light from background stars.

These cold, dark, dense clouds, particularly their interiors, are ideal for the formation of molecules. At first glance, you might say, "Well, wait a minute; some things don't make much sense here. If these clouds have densities that are very low—even the densest ones are very low—first of all, how can they block starlight? If its density is so low, how can that be if they're so much less dense than our atmosphere? The answer is: They're huge; if you have clouds light-years in size, even though the density's low, you can have enough material spread out over a great distance you can block the background starlight. Also, in these individual clouds, even though the densities are low, molecules can form; atoms can run into each other and form molecules. You might say, "That seems to be not a very efficient process given such a low density," and you're right; but the important thing to realize about these dense, dark clouds is that they have dust drains through them, and the dust drains shield the interior of the cloud from the background ultraviolet light from stars that could break up the molecules. So if you have a medium deep inside these clouds where molecules can form, albeit slowly, but you've shut off the main destruction mechanism, given time—and in astronomy, we always have lots of time for things to happen—you build up molecular abundances deep in these cloud cores. That's why astronomers typically refer to dark clouds as molecular clouds, because deep in their cores they have lots of molecules. Indeed, it's an amazing molecular chemistry. We find molecules ranging from formaldehyde to ethyl alcohol; indeed, some molecules have as many as 13 atoms in them.

Particularly why these dense regions of dark molecular clouds are so important is because this is where the slow gravitational contraction begins that gets the process of star formation going. Once these regions begin to contract under their own gravity, their interiors slowly heat up until the core hydrogen fusion begins and a star is born. It takes roughly about 10 million years or so to produce a solar-type star in this manner.

The prototype of the smallest type of molecular cloud, a Bok globule, is Barnard 68; sometimes called the "Black Cloud." Here's a ground-based image covering a sky area equivalent to about seven percent of the full Moon, and when you look at this image the cloud stands out; it's just amazing, it just stares right out at you, this rich Milky Way starfield. It literally looks like a hole punched right into space. The physical dimensions of this cloud: It's about a half light-year across. One of the things you really notice when you look at this, there are no foreground stars; there are none. There are no stars in front of this; it literally looks like a hole. You might look at this and think, "There are just no stars there." No, there's a cloud here, and it's absorbing the light of background stars at optical wavelengths. We know this cloud must be nearby, because if it was very far away we'd see stars in front of it. The fact that we don't mean it has to be reasonably close; and indeed, it's about 500 light-years. In terms of the galaxy and the Sun, 500 light-years, that's pretty close among stars and interstellar clouds.

If you look at this image even more carefully, if you study the edge of the cloud, you see these red stars on the edge of the cloud. If you look farther away from the cloud image, you see stars of all kinds of colors; but why is it just on the edge of the cloud you see just red stars? What's going on there is interstellar dust at work, because interstellar dust, as light passes through it, scatters blue light more than red light; so as the light's coming through a cloud toward you, the blue light's scattered out of the way so that your eye sees the red light. That's what's going on in the edges of this cloud; the dust thins enough so you can begin to see the stars behind the cloud, and you see them appear red. There's a good analogy to this: If you've ever been in a city on a smoggy day and you watch the sunrise, it looked very red, much redder than you could usually imagine a sunrise or a sunset looking; indeed, if you go somewhere else that doesn't have any smog at all, the typical yellowish sun seems to be there arising, or certainly when it's up high in the sky.

The thing to understand at sunset and sunrise: When you look at the sun there, you're looking through a lot of atmosphere on the horizon; so all those dust grains, if you have a smoggy day—and in many respects, smog has characteristics very similar to interstellar dust—so as this sunlight comes on the horizon on a smoggy day, the smog dust basically scatters the blue light out of the way and all you see is red light on the horizon when the sun's there. When the sun is up higher in the sky, or in a place where there is no smog, you see the sun at its normal color. In other words, here's an analogy to how the atmosphere can change the color of the sun in a somewhat similar way that interstellar dust in the galaxy can change the color of a background star. This particular cloud—this Bok globule; the black cloud here—has a total mass of about two solar masses, and studies have not revealed any stars forming in this cloud yet. You might say, "How could we know?" Even though this cloud is dark to us at optical and ultraviolet wavelengths, at infrared wavelengths, infrared light can get through the dust to some extent, so we can probe the interior of this cloud with infrared light. Such studies have been done and no young stars have been found to be forming inside this cloud.

Our next image is that of the Horsehead Nebula. This is a much larger molecular cloud, and much more massive; by more massive, I'm talking about 80,000 solar masses. This particular image of this cloud covers five times the sky area of the full Moon. One first indication, just by looking at this image to see where there is dust, is if you compare the star counts in the bottom half of the image to the top half, you see the bottom half has fewer stars; at some level, the star counts alone, besides any other evidence, give you some indication where there is this dark, dusty cloud. In addition, you see this characteristic "horse head" rising above this dark cloud. To give you a sense of scale here: This is a big horse head; this horse head rises a few light-years above the cloud.

In addition to this characteristic, the reason the horse head stands out here, is you see this red glow—you see this red background glow in places— illuminating partially the surface of this dark cloud. What is this due to? This radiation comes from the hydrogen in the cloud itself, and in the background hydrogen in the more diffuse part, rising in the upper half of the image. How does hydrogen glow in this red wavelength? The way this works is just off the edge of this image is a hot star with a lot of ultraviolet light, and this ultraviolet light is hitting the hydrogen gas in the vicinity of this cloud. As this ultraviolet light hits the hydrogen atoms, it breaks them up into its constituent proton and electron; in a normal hydrogen atom, an electron orbits a proton in the nucleus. These ultraviolet photons come in from this hot star and give the electron enough energy so it gets ionized and removes itself from the proton. So we have here in the vicinity of this nebula electrons and protons, and there are enough of them that some of them eventually recombine; they recombine back into a hydrogen atom.

But the hydrogen atom that they recombine into has a lot of energy; and how does it give up this energy? The electron decides to lower itself in energy level, and in so doing the hydrogen atom will now emit a photon. Hydrogen atoms emit these photons at very specific wavelengths. This leads to the kind of fingerprints we talked about earlier in terms of identifying atoms in interstellar clouds. In the case of hydrogen, the optically brightest emission line is at a wavelength of 656 nanometers—we call this transition H-alpha—and 656 nanometers is in the red part of the spectrum. That's why this nebula appears reddish. The Horsehead is an example of an emission nebula, where a hot star is close enough to ionize part of the cloud and produce the characteristic red glow of abundant hydrogen. If we didn't have such a nearby source of ionizing photons, we wouldn't be seeing any red lighted here at all; the only real evidence at optical wavelengths there was a massive, dense molecular cloud here would be from the star counts effectively. We wouldn't really see that Horsehead as well because this red glow comes from the ionizing of the hydrogen by a nearby star.

The Horsehead Nebula is not alone; it's part of a large complex of molecular clouds in the constellation Orion. Orion is one of the most easily-recognizable constellations in the sky. It's marked by (it's anchored, really) by two really bright stars: Betelgeuse, which is a red supergiant and also goes by the name of Alpha Orionis—the brightest star in the constellation Orion—as we talked about last time stars come by many names; each star has a large number of names, and these stars are so bright they have more names than usual. Here's Betelgeuse, Alpha Orionis, this bright red supergiant on one end of Orion; and down low we have another bright star called Rigel that is a blue supergiant also known as Beta Orionis. Between these two brightest stars that anchor Orion, we have in between them the three stars in Orion's Belt. This is a familiar constellation in wintertime.

The Horsehead Nebula is located near Orion's Belt. Perhaps an even more famous nebula, the Orion Nebula, is located below the Belt toward Rigel. When we look at the Orion Nebula, we're looking at the front part of a 100,000 solar mass molecular cloud, and the illumination of the Orion Nebula comes almost entirely by four really hot stars called the Trapezium Stars deep inside the nebula. This picture of the Orion Nebula: Where we're looking at it, we're seeing something that's 25 light-years across, and it's at a distance of 1500 light-years. The Orion Nebula is so bright you can actually see it with the naked eye; it's the brightest nebula in the sky. If you look at it with the naked eye, it looks as a faint whitish glow.

My first really close up view of the nebula came long ago, and it was really a spectacular one. I was a first-year graduate student at UCLA in Los Angeles, and during my studies that first fall quarter, a visiting astronomer came to time and she was going to be observing with the Mount Wilson 100-inch telescope in the mountains outside Los Angeles during Thanksgiving weekend, and she wanted to know if any one of the first-year students wanted to come along with her. I, of course, raised my hand immediately; and we went up and she was carrying out a research studying interstellar gas clouds and she said, "OK, we're going next to the Orion Nebula, do any of you want to look at it through the eyepiece of the 100-inch telescope?" I said, "Yeah, that sounds like a great idea." Very few people get a chance; this was at a time long ago effectively when astronomers still looked through eyepieces on big telescopes, and to actually see the Orion Nebula with an eyepiece through the 100-inch telescope was something I'll never forget. Amazingly enough, even though I have this 100-inch telescope, I don't see anywhere near the kind of detail one sees with Hubble, and I certainly don't see the kind of colors we see with Hubble; it just was this even brighter white glow and I could see more details, I could resolve apart the Trapezium a little bit better, but it was just an amazing introduction to observational astronomy to get this kind of view of the Orion Nebula with a 100-inch telescope as a first-year graduate student.

If we look at Orion with the Hubble Space Telescope—and it's interesting that Hubble has about the same size primary mirror as the 100-inch telescope at Mount Wilson—what we see with Hubble is a very turbulent nebular structure. We see a mix of dust and this glowing gas, and we also see—if we close in on the Trapezium itself with Hubble—even more details; we see a lot of faint young stars interacting with the gas and dust in this Orion Nebula. There's active star formation going on in the Orion Nebula, but it turns out that most of it is going in the cloud behind the Orion Nebula; it's a very massive molecular cloud, but it's also very dusty, so with optical light we can't really see beyond the Orion Nebula into where the active star formation is going on. We can probe that with infrared and radio telescopes, though, and that's how we know this kind of star formation is going on; but it's too dusty to see with Hubble at optical wavelengths.

Fortunately, the Eagle Nebula provides an optical view inside a large molecular cloud where millions of years of star formation have produced thousands of stars. The Eagle Nebula is about 15 degrees away on the sky from the Sagittarius Star Cloud. This wide-field nebular glow of the Eagle Nebula covers a sky area greater than the full Moon; a 70 light-year width at its distance of 6,500 light-years. In this ground-based image, green denotes H-alpha—so we basically substituted green for red here—and the glow that we see in this nebula is excited by the ultraviolet light from this interior star cluster of something like 50 hot, young stars that were born inside the cloud. The winds and the radiation from these stars have shaped this cloud over millions of years; they've helped to open this 20-light-year-wide window into the interior of this cloud so that we can see into it with Hubble and other ground-based telescopes at optical wavelengths. We can see the sculpted interior of these gas/dust pillars inside the Eagle Nebula.

It was in 1995 that Hubble took our feature close-up image of the prominent gas/dust pillars at the core of the Eagle Nebula and revealed the wealth of illuminating small-scale structure. This particular image is a composite of three different images filtered down on mission lines due to hydrogen atoms, doubly-ionized oxygen ions, and singly-ionized sulfur ions. Each one of these images is a net exposure of about 37 minutes. The colors that were assigned in this image: hydrogen was assigned green, oxygen was assigned blue, and sulfur was assigned red. These particular assignments were done in this manner to really bring out the fine detail in this image. However, the color scheme we have here is not real; the real nebula has a much redder color. As we stare at this image, we see details on the tallest pillar—we see these thin protuberances, little stalks, coming off this tallest pillar—and some of them appear to have stars emerging from the globules at the tips. To give you a sense of scale, the globules here are 400 times the Earth-Sun distance; and to give you a better sense of scale, the tallest pillar is four light-years in length. It's similar to the distance between the Sun and Alpha Centauri; although between the Sun and Alpha Centauri there are no other stars, here we have a case where a four light-year stretch in this particular dust pillar, it's not only very dusty but there's dozens of young stars forming. These globules are essentially star cocoons of gas and dust for the formation of these stars.

How were these pillars sculpted? Nearby hot stars, as their radiation and stellar winds were working on the inside of this molecular cloud, evaporated the less dense material around the pillars, slowly eating it away and uncovering the denser parts; uncovering these dense stalks; uncovering these globules; and beginning to reveal to the outside world (us) that there were stars forming inside. The drawback for these young stars forming is their older siblings are stunning their growth; the radiation and the winds from these older stars—these well-developed stars—is eroding away the very material from which these stars could grow bigger. In some ways, the earlier generation of stars in this cloud is basically limiting how much bigger the following star formation generations could be. Effectively what that's telling us is that in clouds like this, star formation can be self-limiting; in other words, if you have 100,000 solar mass cloud it doesn't mean you're going to get 100,000 solar mass stars to form out of that, the efficiency is much less than that. It's interesting when we look at the other Eagle pillars with Hubble we see similar phenomenon; we see other evidence of these kinds of globules forming and evidence of the dust being eroded away to reveal these dense regions in the cloud.

The Hubble images of other nebulae associated with molecular clouds have been observed and look at, and they are found to reveal small-scale structures similar to those in the Eagle Nebula plus some new clues to the process of star formation. Let's go on a little tour here with these other nebulae and start with the Trifid Nebula. The Trifid Nebula is about 10 degrees on the sky away from the Eagle Nebula. This is an amazing image. If we do a close up, the Hubble close up, on the molecular cloud surface in the Trifid Nebula— this image is about seven light-years across and the Trifid is about 9,000 light-years away—the interior of this nebula looks like the head of a snail poking out from a dark thunderstorm cloud; and this would be a really big snail. If you look at this snail, you see it looks as two eye stalks, and one of the eye stalks has a glowing tip. That's very similar to the globular fingers we saw in the Eagle Nebula. The other eye stalk is a jet of gas; it's not a star forming, it's not a stalk like the other ones we saw, it's a jet of gas coming out of this cloud. It turns out that such jets of gas are a familiar stage in protostar evolution. Let me briefly talk about this.

How does a star form? As a dense cloud region contracts due to a principle we know as the conservation of angular momentum, it rotates faster as it contracts. You've seen this in action. If you've ever watched the Winter Olympics, you've seen an ice skater do a pirouette: As she brings in her arms, she spins faster; and that's what happens in the evolution of a star forming. As it collapses and gets smaller and smaller it also forms an accretion disk around this inner star forming, because as this contraction occurs gravity tends to compress the outer part of it into a disk. Then as this process continues—as the protostar continues to contract slowly—the disk will funnel matter into the central protostar. Sometimes this matter can come in too fast for the protostar to swallow or there can be other factors coming in, and then this matter will be channeled out into bipolar outflows from the forming star, and these outflows can last for over 100,000 years. In addition to these jets, the disk that's left over from star formation may eventually lead to the formation of planets. The bottom line is: Star formation leads not only to stars, it leads to jets, disks, and perhaps planet growth; and they're all potentially observable milestones in the process of star formation.

The Hubble image of the illuminated molecular cloud in the Carina Nebula is another vivid example of star formation in action. This image is absolutely stunning; I mean this looks literally like something out of a fairytale. This image spans about 8 light-years of the 7,500 light-year-distant nebula. Like the HST close-up of the Eagle Nebula, this image, you just look at the details; you see all these dust features all over the place and you see evidence of stars forming in these dust features throughout this region of the nebula. One in particular looks like an elephant trunk—it's amazing how much this dust feature looks just like an elephant trunk—and at its tip is a protostar with a bipolar jet, and one of the jets is moving right off into the nebula and the other jet is creating a bow shock in front of it; and this bow shock is similar to the water-wake you'd see in a speedboat racing across a lake. But this is on a much larger scale: This bow shock here, this is a light-year sized phenomenon we're seeing right here. What's interesting about such mass outflows is they can compress surrounding the gas and perhaps lead to new rounds of star formation. The point is: Star formation processes in a molecular cloud can both limit further star formation by chewing up a lot of the material that stars could use, but it can also lead to compression waves that can get new stars to form.

Besides observing these large-scale nebulae, Hubble has also been successful in directly imaging the individual disks associated with the later stages of star formation. Some of these views give us edge-on pictures of these stars in formation and some of them look like little hamburgers, these edge-on views. You see this dusty disk and some of them show these bipolar jets coming off the protostar at the center of these dusty disks. In addition, Hubble has also obtained face-on images of some of these disks seen against the background nebular glow in the Orion Nebula. Over the course of millions of years, it's expected that such disks will typically evolve into planetary systems. However, as we will discuss in our last lecture, it's much, much harder to find planets around other stars than these large protoplanetary disks. In other words, Hubble can do amazing things for us—we can see these structures in these molecular clouds that show stars in the process of formation; we can see these bipolar jets; we can even see the disks themselves from which planets form—but actually seeing the planets themselves, that's really a hard thing to ask for Hubble.

In summary, the scientific significance of the Hubble Eagle Nebula image is that we can actually see young stars emerging from their cocoons of gas and dust in the densest small-scale structures of a molecular cloud. In combination with Hubble images of the jets and disks in other illuminated molecular clouds along with infrared and radio observations of such clouds, this image has helped astronomers achieve a better understanding of the star formation process. It is clear that the Sun and all of the other stars in the sky were born in a setting somewhat similar to that revealed by HST. It's aesthetically pleasing to know that starbirth is connected with such a glorious vista. It should not be too surprising that the deaths of solar-type stars are also associated with a beautiful sky show. Next time, we'll travel with Hubble to the Cat's Eye Nebula to explore such a stellar demise. Please join us then.

The Cat's Eye Nebula—A Stellar Demise
Lecture 5

The total mass in this outer halo is about half that of the Sun. What it is, is basically the outer-envelope ejecta of the red giant that eventually produced both this and the inner Cat's Eye, and this outer envelope was ejected on the order of 25,000 years ago, based on the speed that we see this outer halo expanding.

The gaseous nebulae that are associated with the births of stars are also associated with the deaths of stars; the expanding ash cloud that attends the death of a solar-type star is a colorful sky object called a **planetary nebula**. The detailed Hubble views of planetary nebulae allow astronomers to look into the exposed innards of dying stars and essentially perform a stellar autopsy. It's the ultraviolet radiation from the dying star that ionizes the atoms in these expanding gas shells that the star has thrown out, and as a result, the gas shells glow. About 3,000 planetary nebulae have been identified.

Among these planetaries, the Cat's Eye Nebula clearly stands out as a top-10 Hubble image. Its structure of rings, bubbles, and knots is complex in detail but beguilingly symmetric in overall appearance. It was discovered in 1786 by the British astronomer Sir William Herschel, who catalogued many galactic nebulae and discovered the

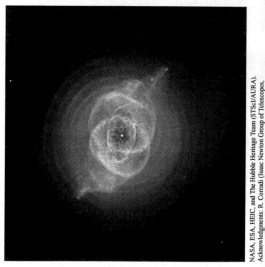

NASA, ESA, HEIC, and The Hubble Heritage Team (STScI/AURA).
Acknowledgments: R. Corradi (Isaac Newton Group of Telescopes,

When we look at the Cat's Eye Nebula, we don't see it as it is now; we see it as it was 3,000 years ago. Its structure of rings, bubbles, and knots is complex in detail but beguilingly symmetric in overall appearance.

NASA, ESA, and The Hubble Heritage Team (STScI/AURA).

The bipolar bubbles and flows, as seen in the Ant Nebula, could possibly be shaped by a strong stellar magnetic field or perhaps by a gravity field of a close companion star.

planet Uranus. Indeed, Herschel came up with the term "planetary nebula" for objects like the Cat's Eye because their sky appearance resembled the small green disk of Uranus.

As we study the Hubble image of the Cat's Eye, we notice several features. We see a series of outer concentric rings of light. These seem to be the result of symmetric mass ejections from the star every 1,500 years for the past 15,000 years. As we get closer to the central star, we see that the Cat's Eye has an axis to it, a bipolar symmetry. As with many Hubble images, this one results in many more questions than it does answers.

The most detailed Hubble image of a planetary nebula in terms of physical length/scale is that of the nearby Helix Nebula. If we look at the Helix with Hubble, we see thousands of cometary knots of gas and dust in the ejecta; each one of these knots is about the size of the solar system. Likewise, if we look at the Eskimo Nebula, we see an inner bubble structure that looks somewhat similar to the Cat's Eye Nebula, and the Spirograph Nebula also has this filamentary bubble-like structure similar to the Eskimo and the Cat's Eye nebulae.

N SA, ESA, and the Hubble SM4 ERO Team.

The Butterfly Nebula has a dusty central torus and fan-shaped outflows of material coming from the central star. Studies of this object show that the central star was probably originally about 5 times the Sun's mass.

It's clear that planetary nebulae exhibit remarkable symmetry and a shape that's circular, elliptical, or bipolar. It's also clear that planetaries exhibit a wide variety in their small-scale structures, whether these are rings, knots, or bipolar structures. Some of the most fascinating planetaries that Hubble has looked at, even though they are very rare, are the bipolar planetaries, such as the Ant Nebula. The Ant Nebula reveals bipolar bubbles and flows; these might be shaped by a strong stellar magnetic field or the gravity field of a close companion star. Another spectacular bipolar planetary observed with the HST is the Butterfly Nebula, also known as the Bug Nebula. When we look at this object, we see that it has a dusty central torus and fan-shaped outflows of material coming from the central star. Studies of this object show that the central star was probably originally about five times the Sun's mass.

The mass of a typical white dwarf is about half that of the Sun, and its size is about that of the Earth.

After the mass loss ends and the planetary nebula moves away, the exposed core left behind at the center of a typical planetary settles down into a **white dwarf**. The central object in the Ring Nebula, for example, is a white dwarf. That's the exposed core of the former **red giant**. The mass of a typical white dwarf is about half that of the Sun, and its size is about that of the Earth. Something that is half a solar mass condensed into something the size of the Earth is obviously going to be very dense, and indeed, a white dwarf is much denser than lead. Although white dwarfs are initially very hot—their temperatures can be more than 20 times the surface temperature of the Sun—they are quite faint because of their small size; remember, they're about the size of the Earth. The brightest star in the sky, Sirius, is only 8.6 light-years away, but its white dwarf companion (the first white dwarf discovered) is far too faint to be seen with the naked eye; indeed, it's 10,000 times fainter than Sirius.

A white dwarf does not generate nuclear fusion energy. It's like a dying coal in a fireplace, and it slowly radiates its heat away. As time goes on, it will change from being a white dwarf to a red dwarf and, in billions of years, become a black dwarf, a dead hunk of matter the size of the Earth but half the mass of the Sun and as black as space. ∎

planetary nebula: The short-lived nebula (lasting for about 50,000 years) that results when a dying red giant blows off the gaseous layers surrounding its core.

red giant: After a solar-type main-sequence star fuses all of its core hydrogen into helium, it evolves into this type of cooler, larger, more luminous star.

white dwarf: The final stage in the evolution of the Sun (and 99 percent of the stars in the Milky Way Galaxy); a compact, Earth-sized object radiating its remnant energy like a dying coal in a fireplace.

Suggested Reading

Balick, "How the Sun will Die."

Balick and Frank, "Shapes and Shaping of Planetary Nebulae."

Kwok, *Cosmic Butterflies*.

Questions to Consider

1. Why are planetary nebulae typically much more symmetric than the nebulae associated with star-forming regions?

2. Is it likely that there are elemental composition differences between the outermost and innermost gas shells in the Cat's Eye Nebula? Why?

The Cat's Eye Nebula—A Stellar Demise
Lecture 5—Transcript

Welcome back to our discussion of the Hubble Space Telescope and its unparalleled view of the cosmic frontier. Last time, we explored the process of star formation through the revealing detail in Hubble's images of the Eagle Nebula and other molecular clouds. It turns out that gaseous nebulae that are associated with the births stars are associated with the deaths of stars; indeed, the life cycle of a star is directly linked to the interstellar medium in that the ashes of dying stars are recycled by the interstellar medium into subsequent generations of new stars. The expanding ash cloud associated with the death of a solar-type star is a colorful sky object called a planetary nebula. Actually they have nothing to do with planets; they were given this name long ago because they typically have a disk-like appearance when viewed on the sky at low resolution with a small telescope. At the extraordinary sky resolution of Hubble, planetary nebulae exhibit a variety of intricate symmetries that put them among the most fascinating objects in the Hubble gallery.

Among these planetaries, the Cat's Eye Nebula clearly stands out to me as a top 10 Hubble image. Its structure of rings, bubbles, and knots is complex in detail but beguilingly symmetric in overall appearance. Whether one sees an eye or a seashell or something else in its colorful patterns, the image is visually stunning. The impact becomes even greater with the realization that such nebulae provide a glimpse of the Sun's distant future. Indeed, the detailed Hubble views of planetary nebulae allow astronomers to look into the exposed innards of a dying star and essentially perform a stellar autopsy. In today's lecture, we're going to discuss the Hubble image of the Cat's Eye Nebula in the context of the evolution of stars like the Sun and the processes involved in their beautiful, albeit transitory, displays that herald the termination of their nuclear energy production.

From our perspective, the Sun is a constant in our lives; it seems to rise and set every day—it not only seems to, it does; it rises and sets every day—and although surface features such as sunspots come and go, we know from just looking at the Sun, its size and luminosity appear to be unchanging. Based on theoretical models and observations of many stars at various stages in their lives, astronomers have a pretty good idea of how the Sun will evolve from its current state of apparent constancy. The way to think about this is to think about the Sun as being in a kind of equilibrium. It's balanced between two forces: gravity, which is trying to squish the sun (and gravity, of course, is ruled by how much mass the sun has); and gas pressure holding the sun up. Where does the heat come from that keeps that gas pressure going? Nuclear energy produced at the core of the sun. These two forces—gas pressure holding up the sun fueled by various nuclear energy processes and gravity trying to squish the sun down—are what compete forever basically in the context of the Sun's evolution.

Let's think about what's going on at the Sun right now. The Sun is being powered through this nuclear fusion of hydrogen nuclei into helium nuclei at the core of the Sun. this can only occur at the core of the Sun, the inner 10 percent of the Sun's radius, because it's only here where the temperatures are high enough—15 million degrees Kelvin—and the pressures are so high that you can take the positively charged protons that are the nuclei of hydrogen atoms and get them to smash together. Remember, if you have two particles of the same charge, they will exert a force to try and prevent you to push them together; so you need a high temperature and a high pressure to slam these protons together and generate the series of reactions that will lead to helium and produce energy. As a result of these reactions, about .7 percent of the mass of the hydrogen atom goes into making energy. You might think, "Well that's not very efficient; that's only .7 percent"; but no, that's extremely efficient. Trying to get energy out of burning oil or gasoline or coal; if the Sun was all made of coal and that's the way the Sun was being powered, it would only last for 5,000 years. But even so, even though hydrogen into helium fusion is very efficient, it still takes 700 million tons of hydrogen fusing into helium every second to account for the Sun's luminosity.

How has 4.6 billion years of fusion changed the composition of the Sun? Originally, the Sun was mostly hydrogen; everywhere in the Sun, from its surface to its center. It was 75 percent hydrogen by mass; about 24 percent of it by mass was helium; and just a tiny fraction, a percent or so, was heavier elements. Now If we look at the composition of the Sun, what we see is the core is now dropped to 35 percent hydrogen by mass, and most of the rest of the core is now helium. But if we look beyond this inner 10 percent that's the core, the outer envelope of the sun, that's still 75 percent hydrogen by mass;

that's important to remember for what we're going to talk about in a few minutes. At the rate that the Sun is burning up the hydrogen right now in its core, it will run out of hydrogen in the core in about 5 billion years.

During the 10 billion years that the Sun is on the main sequence burning hydrogen into helium in its core, it very slowly increases in luminosity. Specifically, the Sun was 30 percent fainter when it was born about 4.6 billion years ago, and it will be about twice as luminous in about 5 billion years. As slowly as the Sun's luminosity changes, it's really hard to explain global warming on Earth being due to the Sun's luminosity changing. Over the time that the Sun's a main sequence star it'll only slowly change, and we won't really notice it that much here on Earth for another billion or 2 billion years. As we look ahead to 5 billion years, when the Sun's core runs out of hydrogen, then things get much more dramatic for the Sun. At that point in time, what will happen: As the core begins to run out of hydrogen, it will begin to slowly contract and heat up; and as it heats up slowly and contracts very slowly, what will happen is eventually the core will heat up a shell around it to a high enough temperature that will ignite hydrogen fusion into helium around the core. What's important about this: This shell around the core will actually produce much more energy than what was in the core before, because the shell's effectively bigger and will burn somewhat hotter as this core contracts and gets very hot. As this happens—as this shell around the core produces a lot of energy—what happens to the rest of the star? The rest of the star all of a sudden sees wow, there's a lot more energy coming down from the inside of the star, and what happens is the gas pressure will increase. As the gas pressure increases, that means the Sun will become much bigger. It has more oomph to beat back gravity, and the Sun can get much, much bigger; indeed, so much bigger its surface will expand beyond the orbit of Venus.

As a result of it getting so much larger, its luminosity will increase tremendously; it will increase by a factor of a thousand. At the same time, it'll cool down somewhat; it'll drop in temperature from about its current 5,800K down to about 3,000K. At this temperature, the Sun, now a red giant, will look very red. Just imagine what it would look like from Earth at this epoch in time, something over 5 billion years from now: The kind of Sun we're used to seeing rising and setting now about a half a degree in the sky, kind of yellowish whitish, this will be instead as a red giant this huge red orb rising and filling a good chunk of the sky; it will be an amazing sight. Unfortunately, of course, there almost certainly will be no one on Earth at that time to see it.

Another thing to recognize about these red giant stars is their outer envelopes, their surfaces, are so far away from the core that the gravity holding is not that tight; it's very tenuous out there. Indeed, red giants lose mass; and the way they lose mass is very interesting, it's tied into those dust grains we talked about last time. In the cool outer envelopes of these red giant stars, the gas condenses out into little dust grains; and then the radiation from the Sun—the radiation pressure from the Sun from the photons the Sun is emitting as a red giant—beat on these grains and push them away, and as the grains move off into space they drag some of the gas with them. This is one important way we make the interstellar dust grains, the stardust, which we find in interstellar clouds.

All this time that the red giant has been getting much bigger and expanding past the orbit of Venus, the core has continued to accrete helium ash from the shell fusion; and thus the core continues to slowly contract and continues to heat up. This added heat increases the shell hydrogen fusion rate, and that makes even more energy come into the Sun. Finally, when this helium core contracts down to a temperature of about 100 million degrees Kelvin, at that temperature the helium nuclei can begin fusing through a series of reactions into carbon nuclei; so all of a sudden the Sun has a new energy source: It has an energy source at the core from helium fusion into carbon in addition to this shell hydrogen fusion into helium energy source. This helps sustain the Sun as a red giant.

Overall, these processes will keep the Sun as a red giant for about one billion years. During this time, it may expand far enough to actually engulf the Earth. What will happen? Your immediate reaction would be to think, "Oh, if the Sun expands to the Earth, poof, the Earth is gone." No, that's not quite what would happen, because remember the outer envelope of this red giant is very tenuous; amazingly tenuous. What would happen actually, as it expands past the orbit of the Earth, the Earth would begin to slow down in its orbit, kind of like a decaying satellite coming down to crash on Earth. The point being right now the Earth is happily going around the Sun in a near-vacuum of solar system space, but as the Sun expands past the Earth's orbit, if it does so, the Earth will slow down and slowly begin its death spiral into the Sun until eventually it falls deep enough to burn up. It's extraordinarily unlikely there will be any life on Earth to see this happen at that point in time, because at the slow pace the Sun is increasing as a main sequence star in terms of its luminosity, it's quite likely to get hot enough in about two billion to three billion years that the oceans themselves will boil on Earth, and at that time complex life, certainly like humans, will be unsustainable on Earth; some bacteria might survive, but at that point any complex life will not be here.

Near the end of its life as a red giant, the Sun will run out of helium in its core. It will not have a carbon core that will then contract, because if it's not producing any nuclear fusion energies, there's nothing to hold back gravity, and so the core begins to contract and heat up some more. But once again what happens, as this core contracts inside a helium-rich shell, it will contract far enough to get hot enough so it will ignite helium fusion in a dense shell around the core. Actually, during its last few million years as a red giant, the Sun will be powered by shell helium fusion in the carbon and a surrounding shell of hydrogen fusion into helium. This is a very important point in time because this rate of shell helium fusion is very sensitive to temperature; just tiny fluctuations in temperature can lead to enormous bursts in energy, and episodic fluctuations in this energy release. As a result, what happens is these thermal pulses will blow off the Sun's outer layers episodically, several points in time; and as a result, in addition to the mass the Sun's already losing as a red giant due to the radiation pressure on dust grains, it will through these processes of these helium shell flashes blow off the rest of the envelope outside the core.

Eventually, the core itself will be exposed to space. The core itself will contract to a point—remember, this is a carbon core—it will keep contracting, but it doesn't contract far enough in the case of a star with a mass like the Sun to start fusing carbon; it doesn't get hot enough; it's halted by something. What it's halted by are the electrons in the core. It turns out you can squeeze electrons only so much before they will exert a pressure to stop you from squeezing them any further. It's like you can only put so many marbles in a bag and tighten that bag so tight; you just can't do it any tighter because the marbles exert a pressure on you. In a crudely similar way, the electrons do the same thing for this carbon core, and thus the core stops contracting and it remains eternally locked between gravity that wants to squish it and these electrons that are fighting it off; and this is the way the Sun eventually dies as a white dwarf. We'll talk more about that at the end of today's lecture.

The key point at this part of the discussion of discussing what happens to the evolution of the Sun—because now we've blown off the outer layers of the Sun and we have this hot, carbon-rich core that's now exposed—since it's hot, it's going to be very bright in the ultraviolet part of the spectrum. That's important, because it's the ultraviolet radiation that can ionize the atoms in these expanding gas shells that are moving away from the star. As a result, as we ionize these atoms—as the white dwarf ionizes the atoms in this expanded gas shell—they will glow in the light of emission lines at select wavelengths in optical, ultraviolet, and other parts of the spectrum. This is what we call a planetary nebula, as in this slide of the Ring Nebula, a prototypical planetary nebula. The lifetime of a planetary is actually quite short, about 50,000 yrs. That's not short, of course, in terms of a human lifetime, but in terms of the kinds of lifetimes we've been talking about in the stages of evolution of a star, this is extraordinarily short. Why is the lifetime so short? Because this gas is expanding away at about 25 kilometers per second; about 25 times the speed of a rifle bullet. That's fast, but it's not as fast as some other things we're going to talk about in the next lecture. But the point is that it's expanding fast enough so as time goes on, it will move away far enough away from the white dwarf at the center—the ultraviolet, bright, white dwarf that, remember, is very small—so far, indeed, that these shells will move away such that the radiation isn't strong enough to cause the atoms to glow.

Thus, there's a finite type of lifetime of a planetary nebula; they simply can't last for much more than 50,000 years because the gas moves too far away to be excited. Although these planetaries are short-lived, what astronomers have found when we look at the galaxy is they're all over the place. About 3,000 planetary nebulae have been identified in the galaxy to date. What that tells us is that most stars must evolve in this way; in order for there to be so many despite their short lifetime, most stars have to evolve this way to produce this large quantity of planetary nebulae. Indeed, what our models tell us is all stars born with less than eight solar masses will eventually become planetary nebulae and evolve somewhat like the Sun, and this amounts to 99 percent of the stars in the sky.

The Cat's Eye Nebula is one of the brightest and most complex planetary nebulae known. It was actually discovered by the British astronomer Sir William Herschel in 1786. He was an avid telescope builder and dedicated observer who catalogued many galactic nebulae and he actually discovered the planet Uranus. Indeed, it was Herschel who came up with the term "planetary nebula" for objects like the Cat's Eye because their sky appearance resembled the small green disk of Uranus. In a wide-field view with a ground-based telescope, the faint filamentary outer halo of the Cat's Eye has a sky diameter that's about one-sixth that of the full Moon. The Cat's eye is at a distance of 3,000 light-years. Think about that a minute: When we look at the Cat's Eye Nebula, we don't see it as it is now, we see it as it was 3,000 years ago. But even so, even on 3,000-year time scales, a planetary would change somewhat; but the point is—I wanted to emphasize this point right here as we talk about the evolution of stars—these light travel times in the galaxy are significant, certainly in terms of a human lifetime and certainly in terms of history on Earth, but the point is in terms of the evolution of individual stars they don't quite matter that much, even in the case of the evolution of a star like the Sun in its lighter stages when it becomes a planetary nebula.

At this distance of the Cat's Eye, this outer halo diameter of the nebula is about 4.4 light-years. In other words, the diameter of this outer part of the nebula is the same as the distance between Alpha Centauri, the nearest star to the Sun, and the Sun. But this outer halo is very faint; it's about 10,000 times fainter than the inner Cat's Eye Nebula. The total mass in this outer halo is about half that of the Sun. What it is, is basically the outer envelope ejecta of the red giant that eventually produced both this and the inner Cat's Eye, and this outer envelope was ejected on the order of 25,000 years ago based on the speed that we see this outer halo expanding.

The Hubble image of the central one light-year of the Cat's Eye Nebula is just absolutely amazing; it reveals all kinds of high-resolution details of the most recent mass loss episodes. This image is actually a composite of images filtered on: one, images filtered on hydrogen and ionized nitrogen emission; and then the other image was filtered on emission from doubly ionized oxygen. The total exposure time of the observations in both of these filters was about one hour. In this particular image, the colors were assigned in the following way: Blue was assigned to these doubly ionized oxygen ions, and reddish orange was assigned to the observations taken through a filter that let in light from hydrogen and singly ionized nitrogen emission. I want to emphasize here: This color scheme is not "real," it's designed for detail. Once again, with the nebular pictures, a lot of times the folks at the Space Telescope Science Institute try to jigger the color scheme to bring out these amazing details in these nebulae.

If you actually looked at the Cat's Eye with a telescope, with your eye just through the eyepiece, it would actually look a little greenish to your eye, unlike the Orion Nebula that, actually if you look at it with the eye through a telescope, it looks kind of just whitish glow; and there's two reasons why the Cat's Eye looks somewhat greenish: Number one, the eye is most sensitive to green at the low light levels we see things in the sky, even with telescopes; but also another key point is what's striking about the Cat's Eye and many other planetaries is they have bright nebular emission lines at wavelengths of 496 and 501 nanometers, and these wavelengths are in the green part of the spectrum. Indeed, it was the Cat's Eye Nebula that the first spectrum of a planetary was taken in 1864. When these lies were seen, the scientists at the time had no clue what they were; they had never been seen anywhere else, on Earth or anyplace else. So what could these lines be due to? Scientists at the time attributed them to a new element they named "nebulium." As time went on, eventually it was realized by 1928 that this was not a new element; these were actually identified as "forbidden" lines of doubly ionized oxygen. These lines only shine when you have very hot, very tenuous gases like planetary nebula; in the kind of environments on Earth, we don't see these kinds of emission lines, and these lines of doubly ionized oxygen can be really, really bright in a lot of planetary nebulae.

As we study this Hubble image of the Cat's Eye, you notice several features. You see a series of outer concentric rings of light. What are these due to? Exactly what's the physical mechanism causing them is a puzzle, but what seems to have happened is there's been periodic, symmetric mass ejections from the star (the central star) for the past 15,000 years every 1,500 years. One possibility of explaining this: It could be to dense pulses of matter in a slow stellar wind as the central object transitions from a red giant to a white dwarf. Remember, as we look at a planet, we're peeling up, it's like tree rings going back in history; as we study, as we peel away each layer going deeper and deeper into the planetary, we're looking back more and more into more recent times. As we get closer to the central star, we're seeing what's happened most recently at the Cat's Eye. What we're seeing is that deep inside, the patterns change, the mass laws pattern changed, about 1,000 years ago; and now there's a much faster wind blowing from the central object and it's interacting with these earlier, more slow-moving shells. It's not understood what's causing this, or what caused the change in the mass loss. In addition, when we look at the Cat's Eye, we see this weird, bipolar symmetry; we see these knots, and the Cat's Eye is on an axis. It has an axis to it; a bipolar symmetry. The amazing thing about this image—and what's always great in astronomy, what's great in science, and what Hubble does more often than you can imagine—these wonderful, detailed images raise a lot more questions than answers.

Since the Cat's Eye is so spectacular, and indeed other planetaries, Hubble has spent a lot of time looking at planetary nebula; they've studied a whole bunch. One of the main reasons you do this is if you look at one like the Cat's Eye and you can't really understand all the details, one way to gain a better understanding is observe a number of them, with the hopes that if you observe a number of them you can look for similarities and differences that could help you understand the mass loss processes of dying red giants. Let's talk about a few of these other planetaries.

The most detailed Hubble image of a planetary nebula in terms of the highest detail, in terms of physical length-scale, is that of the nearby Helix Nebula, which is close by at a distance of 650 light-years. It's big on the sky; it's half a moon in sky diameter. At its distance, that's 3 light-years across. If we look at the Helix with the great detail we can with Hubble, we see thousands of cometary knots of gas and dust in the ejecta; and when you look at this, what's really neat to understand in terms of the scale, each one of these knots is about the size of the solar system. What are they due to; how do we see this? The idea is that these knots came from the earlier ejecta of the envelope of the star; and then as this wind transitioned as the hot core got uncovered, a hot wind and a hot ionization front blew into these outer envelopes and it sculpted away the less-dense regions of gas and dust, leaving behind the dense areas that we see as these knots.

Do we see knots in other planetary nebulae? Yes, we see them in the Eskimo Nebula. The Eskimo Nebula is about eight times further away than the Helix Nebula; and if you look at the Eskimo Nebula, you also see an inner bubble structure that looks somewhat similar to the Cat's Eye Nebula. Another nebula seen with Hubble is the Spirograph Nebula, which is about three times further away than the Helix Nebula. This one is about a tenth of a light-year across, and it also has this filamentary bubble-like structure similar to the Eskimo and the Cat's Eye Nebula. You may wonder: Where are all these names coming from? There is not some single body of astronomers that tries to decide, "What is the name we give each planetary?" Over the course of history, these just gain names by astronomers who study them, publish a paper, and then just say, "OK, we'll call this the Eskimo Nebula." Then if other astronomers like it, they adopt that name and before you know it, 10, 20 years go by and it's the Eskimo Nebula. All these planetaries with different names have a long history—the history can be as long as 100 years or as short as just 20 years—of how these objects were named.

Overall, as we look at this wide variety of planetaries observed by Hubble, it's clear that planetary nebulae exhibit remarkable symmetry and a shape that's either circular, elliptical, or bipolar. It's also clear from the Hubble image gallery that planetaries exhibit a wide variety in their small-scale structures, whether it be rings, or knots, or bipolar structures. What this is telling is there are many factors potentially involved in these differences; in other words, that the evolution of a star from a red giant to a white dwarf in the process of becoming a planetary nebula is a very complicated process and there's lots of physics going on. There could be things varying such as there could be varying speeds and rates of multiple mass-loss episodes. The mass-loss geometry, it doesn't necessarily have to come off spherically; it could come off equatorially off the star, or perhaps in a bipolar nature. What these kinds of variations may be telling us is that some of these planetaries may actually be interacting with a binary star companion. In other words, many stars in the sky actually have companions that orbit them, other stars; so it's possible that some of these planetaries, some of their interesting structures, may be due to another star orbiting it. What we're also seeing in these variations, specifically in the Cat's Eye, even in that short planetary nebula timescale, we see these kinds of characteristics exchange on these short time scales.

Some of the most fascinating planetaries that Hubble has looked at, even though they are very rare, are the bipolar planetaries, like the Ant Nebula; you look at the Ant Nebula and you see these bipolar bubbles and flows, this could possibly be shaped by a strong stellar magnetic field here, or perhaps it's due to a gravity field of a close companion star. Another spectacular bipolar planetary observed with the Hubble Space Telescope is the Butterfly Nebula, which also is known as the Bug Nebula (between the two, I'd rather call it the Butterfly Nebula). When you look at this object, you see it has a dusty central torus and fan-shaped outflows of material coming from the central star. Studies of this object show that the central star was probably originally about five times the Sun's mass.

After the mass-loss ends and the planetary nebula moves away, the exposed core left behind at the center of a typical planetary settles down into this white dwarf. We see this clearly when we look at the Ring Nebula; we see this central object at the center of the Ring Nebula, that's the white dwarf. That's the exposed core of the former star, the former red giant. A typical white dwarf has the mass of about half a Sun, and its size is about that of the Earth. So you stuff half a solar mass into something the size of the Earth, it's going to be very dense; indeed, a white dwarf is much denser than lead. Actually, what you have here is a very dense carbon-rich gas, and it's held up by electron pressure. Understand, this is the way a white dwarf will last for eternity; all eternity, this is the way its structure will last. It'll cool down, as we'll say in a few minutes, but this is how it will last for all eternity. If I went to a white dwarf and I scooped out a teaspoon full and I brought to back to Earth, I'd see that it would weigh a few tons; that's how dense these objects are.

Although white dwarfs are initially very hot—and they can be over 20 times the surface temperature of the Sun—they are quite faint due to their small size; remember, they're about the size of the Earth. The brightest star in the sky, Sirius, is only 8.6 light-years away, but its white dwarf companion is far too faint to be seen with the naked eye; indeed, it's 10,000 times fainter than Sirius. The Sirius companion was the very first white dwarf discovered. I know something about this because it was detected with the very telescope, the 18.5-inch Dearborn telescope, that's in the building that houses my office at Northwestern University. It was detected with the lens of this telescope in 1862. It's really gratifying: This telescope that has so much history and is such an important discovery is used even toady; we use it in our classes, we use it to educate hundreds of students taking our astronomy courses, and we have public nights every Friday night with this telescope, this historic telescope, that's still a fantastic telescope to look at the moon, look at the planets, and our undergraduates can also do research with it even today.

Back to the white dwarf: The thing to remember is a white dwarf does not generate nuclear fusion energy; it's just this thing that's balanced between gravity and electron pressure. Although it stars off hot, it slowly cools; it's like a dying coal in a fireplace, and it slowly radiates its heat away. As time goes on, it will go from being a white dwarf to a red dwarf and eventually become a black dwarf in billions of years, and that is the fate of the Sun; that's how the Sun will end up: as this dead hunk of matter the size of the Earth but half the mass of the Sun and as black as space.

In summary, the wealth of small-scale detail in the Hubble image of the Cat's Eye Nebula has revealed that the planetary nebula stage in the evolution of a solar-like star is much more complicated than simply blowing off the outer stellar envelope shortly after the termination of nuclear energy in the core. The episodic mass loss that produced the concentric rings around the Cat's Eye and the bipolar knot symmetries are all real puzzles. Combined with the wide variety of detail observed in other planetaries by Hubble, it's clear that the astrophysics behind this brief final epoch in the lives of most stars is just as rich and colorful as the images themselves. It turns out that an even more spectacular fate awaits the most massive stars in the galaxy. Next time, we'll voyage with Hubble to the Crab Nebula, the site of a supernova explosion observed on Earth in the year 1054. Please join us then.

The Crab Nebula—A Supernova's Aftermath
Lecture 6

[Betelgeuse is] going to go supernova. It's a red supergiant; it will explode as a Type II supernova, but when? It will happen sometime in the next 100,000 years; we can be certain of that. It could happen within the next 10,000 years. It could happen tonight. Predicting when a given star is going to blow up as a supernova is much harder than even predicting earthquakes, which is very hard to do.

The relatively few stars born with a mass exceeding 8 solar masses are the hottest and the most luminous stars on the main sequence, and they quickly evolve to a spectacular termination within just a few million years. They detonate in a colossal explosion known as a supernova, with a peak luminosity equivalent to a billion Suns. Among **supernova remnants**, none is more famous than the Crab Nebula. Its position in the sky corresponds to the location of a supernova witnessed on Earth in the year 1054. The Hubble image of the Crab certainly looks like the aftermath of a violent explosion; it lacks the graceful symmetry of a planetary nebula. We will explore this Hubble image in the context of its present physical character and the supernova explosion that produced it.

Among supernova remnants, none is more famous than the Crab Nebula. Its position in the sky corresponds to the location of a supernova witnessed on Earth in the year 1054.

Observations and theoretical models show that once a massive star leaves the main sequence, it evolves quite differently than does a star like the Sun. Specifically, as the core of such a star contracts to a temperature and density at which helium fusion can begin, the star expands into a red **supergiant** with a diameter greater than that of the current orbit of Mars around the Sun. After the helium in the massive core runs out, the core continues to contract under gravity, forcing carbon fusion into a heavier element, such as neon, which then fuses into a heavier element and so on to iron, with each succeeding stage of fusion going faster.

NASA, ESA, J. Hester, and A. Loll (Arizona State University).

Stars born with a mass exceeding eight solar masses are the hottest and the most luminous stars; they quickly evolve to a spectacular termination within just a few million years. These stars detonate in a colossal explosion known as a supernova with a peak luminosity equivalent to a billion Suns. Among these supernova remnants, none is more famous than the Crab Nebula.

This star is poised for destruction; indeed, the gravitational collapse of the iron core leads to the destruction of the supergiant in a supernova explosion. Iron does not produce energy through fusion; gravity in the core is essentially unopposed. The collapse is so violent that it smashes the core electrons into the protons, converting the core into almost pure neutron matter and releasing a flood of energy in particles called **neutrinos**. When the resultant neutrino shockwave reaches the star surface, we see a supernova explosion. We call these explosions core-collapse supernovae. They are among the brightest objects we can see in the sky; at maximum brightness, a core-collapse supernova can achieve a luminosity of 1 billion Suns.

The records of naked-eye supernovae date back 2,000 years. Chinese, Japanese, Arabic, and even Native American astronomers noted one that occurred in July 1054. Today, when astronomers point their telescopes at the location where this supernova occurred, they see a filamentary web of gas that is 12 light-years across and expanding at a velocity more than 1,500 times faster than a rifle bullet. This Crab Nebula provides evidence for a very turbulent medium; the Hubble image shows large filaments breaking up into tinier filaments. The inner blue glow is powered by a rapidly rotating **neutron star** called a **pulsar** at its center; it's the collapsed core remnant of the exploded supergiant, an object of pure neutron matter.

The neutron star at the very heart of the Crab supernova remnant pulses at a rate of about 30 times a second, and not just in radio light but also in optical light. The speed of this pulsar spin is the result of the increasing rate of rotation at the core as it contracted. The magnetic axis of this pulsar is not the same as its rotation axis; thus, as the pulsar rotates, the powerful magnetic axis swings around rapidly, beaming high-speed electrons like a searchlight and lighting up the nebula with a blue glow.

Hubble has also imaged other supernova remnants, none more frequently than the site of the 1987 supernova in the Large Magellanic Cloud. Observations clearly show a supernova remnant in formation. In the 1990s, the supernova's shockwave began hitting an inner gas ring cast off thousands of years ago by the pre-supernova supergiant. Shock-heated gas glows at the impact points where the shock hit this inner ring, and the ring is now lit up like a pearl necklace from these impact points. The supernova remnant itself is already a light-year in size since it exploded in 1987.

We are quite fortunate that there is no star anywhere near Earth that is likely to go supernova in at least the next 10,000 years. However, we are overdue for the kind of sky show that our ancestors wondered about when the distant Crab supernova exploded in 1054. Indeed, it's quite likely that one of the distant massive stars observable in the sky tonight has already exploded, and it's only a matter of time before we see this event. ■

neutrino: An elementary particle of very low mass produced in nuclear reactions, such as those in the solar core and Type II supernovae.

neutron star: The collapsed core remnant of a Type II supernova; it typically has a mass of about 1.5 solar masses within its radius of 10 kilometers.

pulsar: A rapidly rotating neutron star.

supergiant: After a massive main-sequence star fuses all of its core hydrogen into helium, it eventually evolves into this type of very large, very luminous, cool star.

supernova remnant: The expanding, nucleosynthetically enriched gaseous remnant of a star that has undergone a supernova explosion.

Suggested Reading

Hester, "The Crab Nebula: An Astrophysical Chimera."

Wheeler, *Cosmic Catastrophes*.

Questions to Consider

1. How could HST observations alone lead us to conclude that the Crab Nebula is a supernova remnant rather than an odd planetary nebula?

2. Why is HST much less likely to discover the next galactic supernova than a ground-based observer?

The Crab Nebula—A Supernova's Aftermath
Lecture 6—Transcript

Welcome back to our tour of the cosmos with the Hubble Space Telescope. Last time, we explored the Sun's distant future through Hubble images of the planetary nebulae that colorfully herald the transformation of all but the most massive stars into white dwarfs. A far different fate awaits the relatively few stars born with a mass exceeding eight solar masses. These massive stars are the hottest and the most luminous stars on the main sequence, and they quickly evolve to a spectacular termination within just a few million years. They detonate in a colossal explosion known as a supernova with a peak luminosity equivalent to a billion Suns. Although these kind of events are rare in the Milky Way, they play an important role in shaping the interstellar medium, stimulating star formation, and seeding the Galaxy with heavy elements. Indeed, most of the oxygen in your body and all of the gold and silver that you own were actually forged in supernova explosions that occurred long before the Earth and the Sun were born. Thus, the dynamically-evolving debris fields of recent supernovae are an enticing target for the high resolution Hubble cameras.

Among these supernova remnants, none is more famous than the Crab Nebula. Its position on the sky corresponds to the location of a supernova witnessed on Earth in the year 1054. The Hubble image of the Crab certainly looks like the aftermath of a violent explosion. There is no mistaking it for the graceful symmetry of a planetary nebula. However, its intricate web of twisted gas filaments has a beauty all its own. Furthermore, the science behind its inner blue glow is as fascinating and cutting-edge as it gets. In today's lecture, we explore the Hubble image of the Crab Nebula in the context of its present physical character and the supernova explosion that produced it.

Observations and theoretical models show that once a massive star leaves the main sequence, it evolves quite differently than does a star like the Sun. Specifically, as the core of such a star contracts to a temperature and density where helium fusion can begin, the star expands into a red supergiant with a diameter greater than that of the current orbit of Mars around the Sun. A good example of a red supergiant is the star Betelgeuse. We talked a little bit about Betelgeuse before; it's the red supergiant you see in the constellation Orion, this bright red star. It's the brightest red supergiant in the sky, partly because it's at a distance of only 500 light-years. But it's a big star; its disk is actually big enough to be resolved with Hubble. This is an important point: It's important to realize that the stars are so far away, even those that are quite big, almost all of them, even with Hubble, are just points of light at great distance; even with Hubble, all it sees is a point of light. But Betelgeuse— Alpha Orionis, the brightest star in Orion—is big enough because its size is comparable to Jupiter's orbit around the Sun; it's an unbelievably huge star. So Hubble can barely resolve Betelgeuse and a few other huge stars, those that are particularly close by.

Speaking of Betelgeuse—or other red super giants—what happens after the helium in the core runs out, the core continues to contract. What's different from this kind of star than a solar-type star is the core is massive enough— the typical core mass is one-and-a-half solar masses or more—and with that core mass, as it contracts, it could contract far enough for carbon fusion into heavier nuclei to occur. What can happen continually with this kind of a massive core at the center of a red supergiant: After all the carbon in the core is fused into a heavier element like neon, then the neon can be fused into a heavier element; and so you get these succeeding stages of fusion going on in the core. But the important point is: Each succeeding stage goes faster, and faster, and faster.

Let me give you an example: Let's take a model of a 25 solar mass star; a star that was born with a mass of 25 solar masses. How does this star evolve? How long does it take? These steps going on as it's a red giant, its core helium fusion to carbon takes about 700,000 years. In its final stage, when you have a core that's over one-and-a-half masses worth of silicon, it will convert that to iron in just one day; that's how fast this fusion goes at these very later stages. Indeed, at this stage, the size of this core is now something like the Earth, and its temperature is 3 billion degrees Kelvin. On this day of its life, if we could cut open such a supergiant, it would have an onion-like deep interior. What we would find in essentially its last hour, it would have a core that's almost entirely iron, and around this iron core we'd see shells of silicon, oxygen, neon, carbon, helium, and hydrogen fusion going on; there are all these energy sources being made going on at the core of such a star, helping to maintain this huge, bloated size of a red supergiant. But I want to emphasize: The energy production is going on deep, deep, deep inside this star; the outer parts of the star have no idea what's going on deep inside, all they know is there's lots of energy coming from down there.

This star is poised for destruction; indeed, the gravitational collapse of the iron core leads to the destruction of the supergiant in a supernova explosion. Why and how does this happen? The reason it happens is because iron does not produce energy through fusion. If you let the core of iron contract and heat up, you can't get to a certain temperature where iron will start fusing into something else. What happens is the core quickly contracts and heats up to a temperature of 5 billion degrees Kelvin. At this temperature what happens is it's so hot in the core, the photons in the core have so much energy, and they can break up the iron nuclei into smaller nuclei. What that does is effectively cool off the core a bit and it collapses even faster; and since there's no other core energy source, gravity is essentially unopposed. What that means is in less than one second, the core collapses from something the size of the Earth to something the size of a city.

This collapse is so violent, it smashes the core electrons into the protons; just slams them right together. As a result, it converts these particles into neutrons; this collapse converts the core into almost pure neutron matter. These reactions of smashing the electrons into the protons also release a flood of energy in particles we call "neutrinos." What's a neutrino? These neutrino particles are very low mass; we sometimes call them "ghost particles" in the sense because they don't interact much with matter at all. Literally, you have billions of neutrinos passing through your body every second from the fusion reactions going on at the core of the Sun; they just don't interact with regular matter much at all. In this particular case, this collapse of the core at the center of a supernova here—what we'll call a "supernova"; this red supergiant collapsing—will produce so much energy in neutrinos, it's an enormous amount coming off this collapsing core.

But now think about the rest of the star: The inner part of the star was happily sitting on top of this Earth-sized, nearly iron core. In the snap of a finger, it's down to the size of the city. What these inner gas layers are going to do is go down fast, and they're going to crash toward this collapsed core at speeds up to 15 percent that of light. What's going on here inside this cataclysm about to happen is you have this enormous wave of neutrinos carrying a tremendous amount of energy coming out of the collapsed core; you have the inner part of the star crashing down into the core; and together the net result of this is essentially this matter bounces off the core, carries fuel partially by this neutrino energy; then sends a shockwave rippling through the rest of the star. The star is so big it takes hours for this shockwave to actually reach the star surface. When it does, we see a supernova explosion.

We call these things a core-collapse or a Type II supernova. They are among the brightest things we can see in the sky when they occur, because at maximum brightness one of these Type II supernovas can achieve a luminosity of one billion Suns. The nearest supernova in modern times occurred in 1987 in the Large Magellanic Cloud. The Large Magellanic Cloud is not like a cloud of gas; it's actually a small satellite galaxy of the Milky Way at a distance of about a 160,000 light-years. Despite its distance, this supernova—Supernova 1987A—was naked-eye bright. The Large Magellanic Cloud can only be seen from the southern skies when you are on Earth, so it's a very interesting target in the southern hemisphere. This particular supernova, you could see it with your eye; you'd go out one night and there was just this large Magellanic Cloud in the sky, and then the next night there's this bright thing in the middle of the Large Magellanic Cloud, and this thing's 160,000 light-years away.

What astronomers could do is they could take a picture of the Large Magellanic Cloud with the supernova, and then what the Large Magellanic Cloud looked like before the supernova; and they could compare the pictures and identify the star that blew up. It turned out the progenitor was a supergiant star, which was consistent with the theory. Also impressive was we actually on Earth detected the neutrinos from this supernova; 160,000 light-years away, we detected the neutrinos from this supernova. The only other extraterrestrial source of neutrinos we've ever detected on Planet Earth is from the Sun. as I said, stars like the Sun produce many, many neutrinos, but nowhere near the abundance that are produced in supernova explosions. The point is: The mere fact we can see neutrinos from the supernova just gives you an idea how many zillions of neutrinos come out of these Type II supernova explosions.

When we look out at the universe and we study other galaxies—and we'll talk more about other galaxies a few lectures from now—we see hundreds of supernovae every year. They indicate—based on the studies of these supernovae—that a Type II supernova occurs in the Milky Way roughly once every 100 years; yet the last one seen widely on Earth was in the year 1604. Based on this, we would seem to be overdue for a bright galactic supernova; but the thing you have to remember is at optical wavelengths, we see only a tiny piece of the Milky Way. There could easily been a supernova sometime over the past 300 years clear on the other side of the galaxy in the galactic disk, and despite how bright and luminous such a supernova would be, there's so much dust in the way we wouldn't have seen it at optical wavelengths. But the good news is, now that we have these neutrino detectors on Earth—and they've been operating now in some form or another for about 40 years—it may well be that our first indication of the next galactic supernova may not be through its optical light, but we may detect its neutrino burst.

Let's go back to our old friend Betelgeuse. It's going to go supernova; it's a red supergiant; it will explode as a Type II supernova, but when? It will happen sometime in the next 100,000 years; we can be certain of that. It could happen within the next 10,000 years. It could happen tonight. Predicting when a given star is going to blow up as a supernova is much harder than even predicting earthquakes, which is very hard to do. The point is: When Betelgeuse goes off as a supernova, it will be brighter than the full Moon. This single star, when it goes off as a supernova, will brighten up—take all the light of the full Moon and imagine compressing that into a single point of light; that will become what Betelgeuse will look like on the sky—and it'll be that bright on the order for a few weeks to a month or so. If we had an even closer star explode as a supernova, it would make night twilight for a few weeks; you could even read by its light at night. It'd be something no one could ignore; even a cursory examination of the night sky, even from a city, we'd say: "What is that bright thing in the sky?" A nearby, bright galactic supernova would be an astonishing thing to see.

This is the kind of event that Hubble would jump on. We refer to these kinds of events—like a supernova, not only in our galaxy but other galaxies; or novae explosions; or other kinds of things we see in the galaxy—as targets of opportunity; things that you can't predict exactly when they're going to happen, but if they do happen you want to jump on them with a telescope and study them because seeing what's actually happening at the moment of the explosion or shortly thereafter is really important to science. It's possible with the Hubble Space Telescope to write a proposal for a targeted opportunity observation. You can write a proposal, and Hubble takes proposals every year, and you can say, "I'd like to observe the next galactic supernova explosion as a targeted opportunity sometime over the next year." Of course, you won't get the observing time unless you actually spell out the science you want to do with those observations; and Hubble will evaluate that and decide whether to award you the time. You will only get the time if the supernova occurs; if it doesn't, you don't get the observing time. But Hubble can only do so many of these; indeed, for Hubble to look at a targeted opportunity target within the space of two days, it can only do that once or twice a year.

You might say, "Why? Why can't Hubble just jump on anything any time?" Because scheduling Hubble is so hard. Remember we talked earlier about Hubble's in low-Earth orbit, so we have this big Earth in the sky; as Hubble orbits the Earth it can't point anywhere near the Earth, it can't point anywhere near the Sun; and remember it moves at the rate of half an hour to go from this direction (180 degrees) to opposite direction. What that means is the way Hubble is scheduled is as the proposals are evaluated every year and it's decided who gets the time and what targets will be observed, they're all put together in big computer schedule; it's kind of like the traveling salesman problem. In other words, how do you minimize the amount of time making trips, because every minute of Hubble time is valuable? You don't want to be going from this target and taking a half hour to go here, and a half hour to go here, etc.; you want to schedule your observations so you go from here to here to here, you waste as little time as possible going from one target to another.

If you have to respond to a targeted opportunity that throws your schedule way out of whack. "Oh gosh, there's a supernova, we have to swing way over here and observe that for a while," and it throws the whole schedule out of whack and Hubble loses lots of time. The amount of time that Hubble can lose by doing one of these is a lot. That's why policy is they only schedule a couple; only one or two of these per year are allowed. They will do more targeted opportunities if you don't need to get there in 2 days but maybe 5–10 days; and some of those can be done if the supernova goes off in another galaxy. But supernovae in particular are such exciting and important events, and relatively rare even beyond the Milky Way, that Hubble takes all the opportunity it can to do good science with these objects.

Let's go back to talking about the supernovae themselves. The records of naked-eye supernovae date back 2,000 years. This shouldn't be too surprising; for people many thousands of years ago, the sky was an important part of their lives. They got used to seeing the constellations in certain places, so anything new was, "Whoa, where did that come from?" A comet would always be, "Wow, where did that comet come from? This is important. What constellation is it in? What does that mean?" In July, 1054, a guest star appeared in the constellation Taurus. This was noted by Chinese, Japanese, Arab, and Native American astronomers, and the Native Americans, we see on one rock in the American Southwest a pictograph indicating that it looks like they saw this event some thousand years ago, and they thought it was so important they made a record of it on the rock; and the reason it was so noticeable, it was 10 times brighter than the planet Venus in the sky, and it's actually visible during the daytime.

When astronomers point their telescopes at the location of this supernova in the year 1054 today, they see a filamentary web of gas that is 12 light-years across and expanding at a velocity over 1,500 times faster than a rifle bullet. The Hubble image of the Crab Nebula provides the highest-resolution view to date of the exploded innards of the supernova progenitor. This image is actually a mosaic of 24 individual images using three different filters. The total exposure time of this image is about 28 hours, so a lot of time was invested in this image. In terms of its area on the sky covered by the Crab Nebula, the sky width is about one-fifth that of the full Moon. When we look at this nebula and we see the different colors in the HST image, the filament colors are chosen to accentuate the light of different elements. Oxygen atoms are bluish in color, doubly ionized oxygenized ions are reddish in color, and singly ionized sulfur ions are greenish in color. Note specifically the color scheme in the Crab Nebula, this is not really what you would see with your eye; it's designed for detail. Also, another key point, the inner blue glow you see in this image is not due to atoms; we'll talk more about that in a few minutes.

If we look close by and study a section of this Crab Nebula image more carefully, we see evidence for a very turbulent medium; we see these big filaments breaking up into tinier filaments. What's going on here is the turbulence associated with this explosion is mixing together oxygen, sulfur, and other newly forged elements in this supernova explosion; and as this nebula expands further into the ambient interstellar medium, these newly forged elements will be mixed with those such that down the road millions, billions of years from now, new stars will be enriched in these heavy elements produced by the Crab Supernova.

What's the source of the energy that excites these atoms and ions in the Crab filaments to glow like we see in this image? It turns out that the Crab Nebula is powered by a rapidly-rotating neutron star called a "pulsar" at its center. What's a neutron star? It's the collapsed core remnant of the exploded supergiant. Remember we talked about this thing that collapsed from something the size of the Earth to something the size of a city? We smashed the protons into the electrons and we created an object of pure neutron matter? That's what a neutron star is. What holds this thing together? Just like a white dwarf balances gravity with electron pressure, here you can only squeeze so many neutrons together before they all exert a pressure to hold off gravity; and that's what holds up gravity in a neutron star. Neutron pressure holds up a neutron star against gravity trying to force it down. This is an unbelievably weird object; it has one-and-a-half solar masses stuffed inside a ball with a diameter of 25K. This is much, much, much denser than a white dwarf; if you took a teaspoonful of a neutron star and brought it back to Earth, it would weigh a billion tons. This is so far beyond anyone's imagination it's hard to even believe such things exist. Indeed, when the idea of neutron stars first came up in the 1930s, most folks looked at it and said, "That's crazy"; even if it did exist, though, we'd never find it because they're so small; how could you find such a thing?

They were found, and they were found in the forms of pulsars. These pulsars were first observed in the 1960s, and the way pulsars were found: People were doing radio observations in the sky, and they found in certain directions they were detecting sources that were beeping radio light—it would beep radio light, "beep, beep, beep, beep" like every second; every second there'd be a beep from a particular source of radio light—and the pulses were coming at an amazing regularity; the periods were better than any watch on Earth. What were these things? When they were first discovered, some folks thought, "Well this could be the first sign of extraterrestrial life; these are alien civilizations sending us a message." Why the aliens would be sending us pulses that beep at the same monotonous rate is not clear; but the very first pulsar was designated LGM1, where the "LGM" stood for "Little Green Men." But it turned out as more studies were made, as more and more of these pulsars were found, it didn't make much sense that this was probably an extraterrestrial civilization sending us messages, it really urged astronomers to try to think of some natural explanation. Then, at the very heart of the Crab Supernova remnant, a pulsar was found; and the Crab pulsar was an amazing pulsar, because it was found to pulse not just once a second but something like 30 times a second; and the Crab pulsar was pulsing not just in radio light, but in optical life. You couldn't imagine a white dwarf spinning, or pulsing, or doing anything 30 times a second; it couldn't be a white dwarf, it actually had to be a neutron star, and pulsars were interpreted as rapidly rotating neutron stars.

Why does the Crab pulsar spin so fast? It's due to the principle we talked about in the formation of stars: this idea of conservation of angular momentum. As the core of the supergiant contracts from something the size of the Earth to something the size of a city, it gets much, much, much smaller; and any rotation the core had when it's the size of the Earth gets much, much, much faster when it contracts down to the size of the city, just like an ice skater when she's doing a pirouette on the ice: she brings her arms in, she spins much faster. Also, you would expect that such an object would have a very intense magnetic field; as also as you shrink that core from something the size of the Earth to something the size of a city, its magnetic field would also be much more intense. Indeed, the magnetic field of a pulsar is typically 10 billion times stronger than that of a bar magnet. It's also important to understand that the magnetic axis of this pulsar is not necessarily the same as the rotation axis of the pulsar; so as a result, the magnetic axis can swing around really fast as the pulsar rotates very fast.

What happens near the pulsar's magnetic poles is electrons go spiraling along these magnetic field lines. As they spiral along these magnetic field lines to near the speed of light in the vicinity of the poles of the pulsar, they emit a particular kind of radiation called synchrotron radiation, and this radiation comes beaming out of the polar regions of the pulsar. As this pulsar spins, these beams of synchrotron radiation sweep round and round into space, just like a lighthouse; and if you're in a distant star that's in the line of sight for one of these beams, you're going to see a pulse of light—radio light, or in the case of the Crab pulsar, optical light—every time it sweeps around. If you're not in the beam, you won't see it. These beams are what produce these radiation pulses that are what's observed. What the pulsar also does is its magnetic field is so intense and so vast it pumps these high-speed electrons right into the nebula. It has a pulsar wind, a wind of these particles, coming right off the pulsar; and what's really neat is you look at the Hubble images taken over weeks to months and make a little movie out of them, you can actually see the dynamic variations in the Inner Crab Nebula as these pulsar winds interact with the nebula. This inner blue glow I mentioned earlier can also be explained by this electron synchrotron radiation: As this energy flows from the magnetic field into the nebula, it radiates and causes the blue glow; furthermore, this energy also excites the atoms and ions in the filaments.

The Crab pulsar is the engine driving all of this activity—it's the engine driving the blue glow, the wind nebula, the glowing we see in some of the filaments—and it's currently pumping out the energy equivalent of 100,000 Suns. But the energy has to be coming from somewhere, and indeed where it's coming from is this rotation of the pulsar, so we can actually observe its slowing as the energy is drained; as we monitor the period of the Crab pulsar pulsations, we can actually see it slowly going down with time. That's why the Crab Nebula is so fascinating; not just because it's an amazing-looking nebula, but for the beast at the center of the Crab Nebula and the amazing power it has to fuel the nebula.

Hubble has also imaged a variety of supernova remnants younger and older than the Crab. None has been observed more frequently than the site of the 1987 supernova in the Large Magellanic Cloud. There have been multi-epoch Hubble observations of Supernova 1987A, and they clearly show a supernova remnant in formation. The largest gas rings in this image are due to mass loss at an earlier time for the pre-supernova supergiant, and these rings were actually mass-ejected thousands of years before the supernova. What's neat is the supernova's shock began hitting the inner gas ring in the 1990s. This shock-heated gas glows at the impact points where the shock hits this inner ring, and as we watch it evolve over time we now see that the ring is now lit up like a pearl necklace from these impact points. At the very heart of this is the exploded star remnant itself, and the supernova remnant itself at the heart of this 87A remnant is expanding 10,000 times faster than a rifle bullet and it's already a light-year in size since it exploded in 1987. The elongation we see here indicates that the explosion wasn't spherical. I have to add there: There's no evidence yet of a central pulsar in 1987A, although people continue to search for such one developing; and when you look at this image and you see the detail we can see, remember this particular supernova remnant is 20 times further away than the Crab Nebula. Only Hubble can see such detail at that kind of a distance.

As supernova remnants age, they fragment, fade, and diffuse into the interstellar medium. Cassiopeia A is a 330 year old supernova remnant and is one good example of such a supernova remnant. The supernova was not widely seen due to dust and distance; it's a distance of about 10,000 light-years, and at that distance there is a lot of dust in the sightline and so it's possible one or two people on Earth may have seen it, but for the most part it was not widely seen. As we look at this image, we see a broken filamentary shell; the shell's about 13 light-years across. We can see another supernova remnant was actually observed by many folks in the year 1604 that was called Kepler's supernova. This is also a very distant supernova remnant; it's at a distance of about 13,000 light-years. This remnant is now 14 light-years across. This image shows what the supernova remnant looks like in different wavelength regions. The optical image shows filaments of shocked, dense gas; the X-ray image shows a remnant bubble of the hot, tenuous gas produced by this supernova explosion; and the infrared image shows the emissions from the dust, because what a supernova explosion, its shockwave, going off into the interstellar medium, it can sweep up the dust grains. As it piles up these dust grains next to each other, they can heat up and actually glow in the infrared part of the spectrum.

Another really fascinating supernova remnant is the Veil Nebula, and it's fascinating because it's just an old nebula, an old remnant, that actually was a result of a supernova that exploded 5,000 years ago. Because it's so old, it covers a big chunk on the sky; it literally covers three degrees on the sky, six full Moon widths. Since this supernova's exploded 5,000 years ago, it's just been steadily moving out into space; and when we look at the Hubble image—the Hubble image covers just a tiny section of it, just about a light-year across—you can see even today the amazing filamentary structure in this supernova remnant; and what you're seeing here is you're seeing detailed small-scale structure in the shock-heated gas. The importance of these supernova remnants is the remnant shock can actually trigger star formation. If you have one of these supernova go off near a giant molecular cloud, as the shock front goes flying into the cloud it will disrupt a lot of the grains and heat up the gas and re-disrupt the front edge of a cloud, but as the shock propagates further deep into the dense part of the cloud, it may possibly kick start gravitational contraction and get stars forming. I find this very ironic: The death of one star in a massive supernova explosion can actually lead to the birth of many other stars if it happens to be close to a giant molecular cloud.

In summary, the Hubble image of the Crab Nebula is a snapshot in time of a process through which the deaths of the most massive stars in the galaxy seed the Milky Way with the elements of life and stimulate the formation of new stars. It is also an exhibition of nature at its most violent, with extremes of velocity, density, and energy that are far beyond human experience and push the limits of our scientific understanding. We are quite fortunate that there is no star anywhere near Earth that is likely to go supernova in at least the next 10,000 years. However, we are overdue for the kind of sky show that our ancestors wondered about when the distant Crab supernova exploded in 1054. Indeed, it's quite likely that one of the distant massive stars observable in the sky tonight has already exploded and it's only a matter of time before its light reaches Earth.

Next time, we'll gaze beyond the Milky Way with Hubble and begin to explore the universe of galaxies with a visit to the majestic Sombrero Galaxy. Please join us then.

The Sombrero Galaxy—An Island Universe
Lecture 7

The galaxies nearest our Milky Way form a cluster called the Local Group. The Local Group consists of 40 or so galaxies. It's dominated by three big spirals: the Milky Way; Andromeda; and M33, the Triangulum Galaxy. The rest of the galaxies in the Local Group are dwarf ellipticals, irregulars, and dwarf spheroidals.

In this lecture, we discuss Hubble images of galaxies in the context of the discovery, characteristics, and local distribution of these "island universes," a term originally coined by the German philosopher Immanuel Kant in 1755. Galaxies were known to astronomers long before their recognition as individual star systems far from the Milky Way. In the 1770s, the French astronomer Charles Messier catalogued more than 100

NASA, ESA, S. Beckwith (STScI), and The Hubble Heritage Team STScI/AURA).

Spiral galaxies are gas-rich systems of young and old stars. The Whirlpool Galaxy (M51) is a classic face-on Sc spiral. It has a kind of small bulge with large, loosely wound arms. An Sa spiral would have a bigger bulge and more tightly wound arms.

fuzzy objects in the sky that were fixed with respect to the stars. The Crab supernova remnant (known as M1) was the first object on Messier's list. Using Messier's catalogue, scientists noted that fuzzy objects above and below the plane of the Milky Way showed much more symmetry than ones in the plane, with some being spherical and others globular. Another class of fuzzy objects exhibited a spiral symmetry, and these spiral nebulae were a complete mystery. How far away they were continued to be a mystery until the early 1920s.

A key part of the solution to the spiral nebulae problem came from the work of Henrietta Leavitt in the early 1900s. Leavitt noticed that there is a relationship between the period of a **Cepheid variable** star and its luminosity; specifically, she found that the Cepheids that have longer periods are more luminous. This Cepheid period-luminosity relationship enables astronomers to derive the distance to any star group that has a Cepheid in it. The man who used the Cepheid period-luminosity relation to solve the spiral nebulae problem and begin the exploration of the extragalactic universe was Edwin Hubble, the most famous astronomer of the 20th century and the namesake of the space telescope.

The man who used the Cepheid period-luminosity relation to solve the spiral nebulae problem and begin the exploration of the extragalactic universe was Edwin Hubble, the most famous astronomer of the 20th century and the namesake of the space telescope.

Hubble had the good fortune to join the Mount Wilson Observatory essentially at the same time that the new 100-inch telescope came online in 1919. Hubble set his sights on working on the spiral nebula problem. He focused on the Great Spiral Nebula in Andromeda (M31)—what we today call the Andromeda Galaxy— and he eventually realized he could identify Cepheid variables in M31. As a result, he showed that the Andromeda Galaxy is much further away than the size of the Milky Way itself. Essentially, what Hubble showed is that spiral nebulae are island universes. He transformed our understanding of the size of the universe.

The Sombrero Galaxy (M104) is one of the 20 brightest galaxies in the sky. The Hubble image of the Sombrero Galaxy provides a spectacular close-up view of this nearly edge-on spiral, with its pronounced central bulge and tightly wound spiral arms. The very bright nucleus harbors a **black hole** of a billion solar masses. We know this because the spectrographs on the HST determined that the central stars in this galaxy are rotating rapidly around something that must be very small and very massive, and only a black hole fits that description. A black hole is essentially an object that's so massive and so compact that its gravity prevents light from escaping. One of the key discoveries made from studying galaxies with Hubble and with ground-based telescopes is that such supermassive black holes are common at the centers of large galaxies. The Milky Way also has a massive black hole at its center.

Today, using the space telescope that bears Hubble's name, astronomers are studying nearby galaxies, such as the Sombrero, in unprecedented detail and charting ever more distant galaxies. By studying these galaxies in detail, we're acquiring new knowledge about supermassive black holes and star formation. ■

Important Terms

black hole: A region of severely curved space around a collapsed stellar core where not even light can escape.

Cepheid variable: A type of pulsating star with a period-luminosity relation that is useful in determining distances to the star's host galaxy.

Suggested Reading

Bartusiak, *The Day We Found the Universe.*

Christianson, *Edwin Hubble.*

Waller and Hodge, *Galaxies and the Cosmic Frontier.*

1. Imagine that the Earth and HST were relocated to a position in the inner disk of the Sombrero Galaxy. How would the Sombrero galactic halo appear from this position? How would the Milky Way Galaxy appear to HST?

2. Suppose that the Sombrero Galaxy was the nearest galaxy to the Milky Way at its distance of 28 million light-years. Would Henrietta Leavitt have discovered the Cepheid period-luminosity relation? Would Edwin Hubble have solved the spiral nebula problem?

The Sombrero Galaxy—An Island Universe
Lecture 7—Transcript

Welcome back to our discussion of the Hubble Space Telescope and its most spectacular cosmic images. Last time, we explored the aftermath of a supernova explosion through Hubble's detailed view of the Crab Nebula. Although such supernovae occur only once every 100 years in our Milky Way, hundreds are observed every year by astronomers in other galaxies far beyond the Milky Way.

The idea of a vast universe containing many billions of star systems like the Milky Way is certainly commonplace today. However, this idea became firmly established only 90 years ago through the hard work of a few astronomers with key insights and new instrumentation. Advances in our understanding of the universe and of galaxies continues today with Hubble discoveries ranging from the gravitational lensing of distant galaxies to the accelerating expansion of the universe. Before we examine these important large-scale Hubble achievements in the lectures ahead, it is worthwhile to spend some time discussing the galaxies in the local universe and the discovery of their nature.

Among the Hubble views of nearby galaxies, none is more striking than the Sombrero Galaxy. The symmetry, glowing halo, and central dust lane of this edge-on spiral easily makes it one of my top 10 Hubble images. It makes my list more for its beauty than its current cutting-edge science, but the image nicely captures the idea of a galaxy as an "island universe," a term that was originally coined by the German philosopher Immanuel Kant in 1755. Located 28 million light-years beyond the foreground galactic stars in this image, the Sombrero and its many billions of stars are millions of light-years away from any similar-sized galaxy. In today's lecture, we're going to discuss the Hubble image of the Sombrero in the context of the discovery, characteristics, and local distribution of these island universes.

It's important to recognize that the Sombrero Galaxy and other galaxies were known to astronomers long before their recognition as individual star systems far from the Milky Way. The story begins with the French astronomer Charles Messier in the 1770s. What Messier wanted to do was find new comets—this was always an exciting thing; the public was always fascinated to learn about new comets—and Messier had a small telescope and he would gaze, looking up at the sky, searching for comets. You recall when we talked about Comet Shoemaker-Levy 9: A comet in terms of the sky is essentially this fuzzy object that moves with respect to the stars. Essentially what Messier was looking for was a fuzzy nebulous object moving with respect to the stars. But as Messier surveyed the sky with his small telescope, he noticed there were other fuzzy objects on the sky that were fixed with respect to the stars. These were the nebula, like the Orion Nebula would be a fuzzy object fixed relative to the stars noted by Messier.

So Messier began to catalog all these nebulous objects through his small telescope across the sky so he wouldn't confuse them with his potential comets, and in so doing he catalogued over 100 such objects. Many of these fuzzy objects were found in the disk of the Milky Way; the Milky Way across the sky. The ones in the Milky Way, many of them were kind of amorphous in appearance; they weren't that symmetrical. Those objects, we know now for the most part, are essentially gaseous nebulae and some star clusters. Indeed, some of the objects we've talked about in the course so far are on Messier's list: The Crab supernova remnant was the 1st object on Messier's list (that's why the Crab Nebula is sometimes called Messier1 or M1); and the Eagle Nebulae, where we talked about stars forming and we saw this wonderful Hubble picture earlier in the course of the cocoons of star formation at the heart of the Eagle Nebula, that was the 16th object on Messier's list, or M16.

In addition to these fuzzy objects along the plane of the galaxy, Messier also found some fuzzy objects above and below the disk of the Milky Way. Some of these turned out to be globular star clusters; we talked about these earlier in the course. For example, M80 we talked about earlier in our course, the Globular M80, was the 80th object on Messier's list. The interesting thing about the fuzzy objects above and below the plane, they tended to show much more symmetry than the ones in the plane of the Milky Way; and indeed spherical ones, some of them were globular clusters. But there was another class that is very interesting that exhibited a spiral symmetry, these "spiral nebulae"; and the origin of these spiral nebulae was a complete mystery. What were they? Most folks out there associated with the Milky Way, and the possibilities included star clusters—that these were distant star clusters so far away that couldn't really be resolved with a small telescope and that's why they looked fuzzy on the sky—or they could be gaseous nebulae like the ones we see in the plane of the Milky Way. But why did they have this spiral pattern? Indeed, some suggested that these spiral nebulae might actually be stars in the process of formation; people were seeing gas spiraling in to a forming star. But there was an entirely different point of view: Maybe these things were distant island universes far, far way; galaxies just like the Milky Way, but at such great distance they were just smudges on the sky, spiral smudges on the sky.

This was an amazing difference of opinion, right? If these objects were part of the Milky Way, then the entire universe was indeed on the order of a few hundred thousand light-years in size. But on the other hand, if these things were Milky Way's so far away they were just a distant, tiny smudge on the sky, then the universe was much, much more vast. This was an important thing to resolve, but it was very hard to resolve for one fundamental reason: We didn't know the distances to these spiral nebulae, and until we figured this out we couldn't resolve the debate; and indeed, this debate continued until the early 1920s.

A key part of the solution to the spiral nebulae problem came from the work of Henrietta Leavitt in the early 1900s. She was one of the first women to break through and make an impact into what was until that time the completely male-dominated field of U.S. astronomy. Leavitt was among a group of 40 women hired by Harvard astronomer Edward Pickering at the turn of the 20th century to analyze the stellar images in a photographic survey made with a telescope in both in the northern and the southern hemisphere; to survey the entire sky, both what you can see from the northern hemisphere of the Earth and the southern hemisphere of the Earth and map all the stars that could be seen and do all kinds of studies of them: measure their positions, measure their brightnesses, measure their colors. The women that Pickering hired for the most part had college science degrees, and these women at the time in society had a very difficult time finding work that would allow them to apply the knowledge they'd gained from their studies; but this was very attractive work, to be involved in such an important project. The drawback is that they didn't get paid very well; they typically only made like a quarter an hour. But this was an opportunity to do science, and so they began to work on this stuff. It wasn't very glamorous work, but it was tremendously important; indeed, it turns out many of the fundamental aspects of stellar astrophysics that were built up later in the 20th century are based in some measure on the work that these women did working for Pickering.

Among these women, Leavitt became an expert in stellar photometry. What stellar photometry is all about is measuring the brightnesses from these photographic plates. That's not as easy as it sounds; it's somewhat easier today when we have these electronic detectors counting up photons, sending them to a computer, and reading out the results; that's pretty straightforward. But when you're trying to figure out how bright a star is based on how big a little dot it makes on a photographic plate, that not only takes skill but eventually that's almost like an art; and Leavitt was a true master at this, the ability to look at these photographic plates and work out how bright the stars are. Let me emphasize once again: This is very hard work: just looking at one photographic plate after another, one after another, and counting literally hundreds to thousands of stars and measuring them in detail on each photographic plate; very tedious, time-consuming work. But Leavitt became so expert at this one of the other task she was asked to do was look for variable stars, look for stars whose brightness varied over time, because as these photographic surveys were done of the sky, pictures were not just taken one time of a particular sky region, but multiple pictures were taken at different epochs so it could actually be seen if things were changing over time.

A key focus in particular of the southern sky survey was the Magellanic star clouds, both the large and the small Magellanic clouds; here we have pictured the small Magellanic cloud. At the time, the small and large Magellanic clouds were thought to be distant star clusters in the Milky Way; so Leavitt began to focus on looking for variable stars in these Magellanic clouds, just studying them all the time, looking for variable stars. Originally, when she first began to do this work, only a few dozen of these variable stars had been found in the Magellanic clouds. By 1907, Leavitt had found 1,777 variable stars; and among these variable stars, a group of 25 very luminous ones in the small Magellanic cloud stood out. We now know these 25 that Leavitt found to be Cepheid variable stars. They have this name; they're named after the prototype star, Delta Cephei, which is the fourth brightest star in the constellation Cepheus. What a Cepheid variable is, is a star that pulsates; it's a super giant star that pulsates, and it does this because its thermostat's essentially a little out of whack. When the Cepheid is big, it's more luminous than when it's smaller. As it pulsates from big to small and big to small, it goes from bright to faint and bright to faint. It turns out the periods of these Cepheids range from a few days to a few months.

The amazing breakthrough that Leavitt made is she noticed that there is a relationship between the period of a Cepheid and its luminosity; specifically, what she found was that the Cepheids that have the longer periods are more luminous. She didn't have to worry too much in this case about different distances of the Cepheids between us and the small Magellanic cloud because the cloud was essentially at a very large distance, and all these Cepheids were essentially at that same distance. This was truly a relationship between the period of a Cepheid and its intrinsic luminosity. Leavitt published this landmark result in 1912; and after this Cepheid period-luminosity relationship became calibrated, it was a key tool for astronomers. Let me explain this, because once this was understood—this relationship between the period and luminosity of a Cepheid—all one has to do is when studying a distant star cluster, one looks at varying brightness. If you find such a star and you thereby determine it's a Cepheid variable, you measure its period; then you go to this Cepheid period-luminosity relationship that Leavitt had established, you look at the line and say, "Oh, for this period, this Cepheid should have this luminosity"; then you measure the mean brightness of the Cepheid and you get the distance to the Cepheid and thus the star cluster that it's in. Thus, with the Cepheid period-luminosity relationship, you could derive the distance to any star group that had a Cepheid in it. Unfortunately, Leavitt passed away in 1921 due to cancer at the young age of 53 before the acclaim that would follow from this wonderful discovery could accrue to her; and she actually unfortunately did not live long enough to see how her work would open the universe.

The man who utilized the Cepheid period-luminosity relation to solve the spiral nebulae problem and begin the exploration of the extragalactic universe that followed was Edwin Hubble. Through this work, Hubble became the most famous astronomer of the 20th century; indeed, NASA named the Space Telescope after Hubble well before launch in 1983 in recognition of his accomplishments and the telescope's goal of measuring the most distant objects in the universe. What I'm going to do in the rest of this lecture and the next lecture in discussing Hubble is try not to get confusion between Hubble the man and Hubble the telescope; so I'll continue to refer to "Hubble" as the man and I'll try to refer to the telescope as "HST."

Edwin Hubble was born in 1889 in Missouri. He was a superb student and a superb athlete. He went to the University of Chicago as an undergraduate, and he played on the basketball team there that won several national championships in the early 1900s. He then won a Rhodes Scholarship and went on to Oxford in England, where he studied to become a lawyer. But it turned out his first love, his first true love, was really astronomy; and when he came back to the United States, he decided to enroll back at the University of Chicago to pursue his Ph.D. in Astronomy, which he earned in 1917. He then went off to serve in World War I, and when he came back to the States he had a position waiting for him at the Mount Wilson Observatory in California. The Mount Wilson Observatory was headquartered in Pasadena, California—part of Los Angeles effectively—and the observatory itself, the telescopes, were in the mountains above Los Angeles and Pasadena.

Hubble had the good fortune to join the Mount Wilson Observatory essentially at the same time that the new 100-inch telescope came online in 1919. This was the forefront observational facility on the planet, and he had the good fortune to be part of the observatory where he could get guaranteed time with this wonderful instrument. The thing that Hubble set his sights on was working on the spiral nebula problem. He began to survey the spiral nebulae and other nebulae above and below the plane of the Milky Way, trying to understand what the nature of these things was. In particular, he focused on the Great Spiral Nebula in Andromeda, the 31st object on Messier's list (or M31) and what we today call the Andromeda Galaxy. With the 100-inch telescope, as Hubble spanned across the Great Spiral Nebula in Andromeda, he found he could resolve parts of it into individual stars; and as he studied it constantly, he eventually realized he could identify Cepheid variables in M31. Here's a slide, essentially one of the photographic plates that Hubble actually took at Mount Wilson, where he's indentifying some of these Cepheid variables. He first thought some of them might be novae explosions in Andromeda, but then he realized as he followed them over time, they were actually Cepheid variables; and thus if he could identify them as Cepheid variables, he could apply Leavitt's period-luminosity relation and thereby find the distance to one of these spiral nebula. This is something no one could do without the power of a 100-inch telescope like the one at Mount Wilson. Then Hubble published a paper in 1925, a very important paper, where he showed that the Andromeda Galaxy, based on these Cepheid distances, is much further away than the size of the Milky Way itself. Essentially, what Hubble showed is that spiral nebulae are island universes. Indeed, today, at its distance of 2.5 million light-years, we now recognize the Andromeda Galaxy as the nearest spiral galaxy among the many billions of galaxies in a universe spanning billions of light-years.

Hubble did this. Think about the work of this man: He essentially, building on the work of others like Henrietta Leavitt and other scientists in the past, he pulled all this together with new technology—this 100-inch telescope—and he dramatically changed the face of the universe, something 100,000 or so light-years across like the Milky Way; taking a universe that many thought that was it to something that was much more big and large, billions of light-years, many billions of galaxies. Hubble transformed our perspective of the universe in terms of its size.

But Hubble didn't stop there; he continued to study these spiral nebulae. He accumulated photos of many of these galaxies, as they were then known; and as he did he noticed there was an amazing variety among the characteristics of these galaxies, and he developed a "tuning-fork" classification scheme based on their elliptical or spiral appearances. When you look at this diagram, it's not like the HR diagram; you shouldn't think of it as an evolutionary sequence. It's just kind of a handy way to catalog the different types of galaxies that Hubble saw. It turns out that 90 percent of the galaxies within our local universe—which is about 100 million light-years—are either elliptical or spiral galaxies. The rest are irregular galaxies like the Magellanic Clouds, the large and small Magellanic Clouds, which orbit our Milky Way.

Among these galaxies, ellipticals tend to be gas-poor systems of old stars. In some ways, they're similar to globular clusters in that there are no young, blue stars in these elliptical galaxies typically, they have very little interstellar matter, and there's essentially no star formation going on in these galaxies. Their shapes range from circular on the sky (E0's) to cigar-like (E7's). There's a tremendous range in size among these elliptical galaxies. Most ellipticals are only a few thousand light-years across, maybe only 1,000 light-years across. These are the dwarf ellipticals. Two good examples of dwarf ellipticals are two galaxies that orbit M31, two satellites of the Andromeda Galaxy. The largest ellipticals can be much, much bigger; they can be several times the size of the Milky Way. Where we find these supergiant elliptical galaxies is typically near the centers of clusters of galaxies, particularly rich clusters of galaxies; and indeed one way these galaxies get so big—these supergiant ellipticals at the centers of clusters of galaxies—is they swallow up some of their companion galaxies who emerge with them over time. We'll talk more about emerging galaxies later in the course.

Spiral galaxies, on the other hand, are gas-rich systems of young and old stars, and in the tuning fork diagram they are essentially classified by the shapes and the sizes of their central bulges. The Whirlpool Galaxy—M51, the 51st object on Messier's list—is a classic face-on Sc spiral. Here's a wonderful HST image of this galaxy. When you stare at M51, you see it has a kind of a small bulge with kind of large, loosely-wound arms; that's a characteristic of a Sc spiral. A Sa spiral would have a bigger bulge and more tightly wound arms. When you look at the Whirlpool, you see this image, you see the dust, and emission nebulae, and star clusters; you have to think when you look at the Whirlpool in the sense that the Milky Way doesn't look exactly like this but it looks something like this. We don't get a view like this of our own galaxy, we're deep inside this; but when you look at the Whirlpool you can get something of an idea, a little bit of how the Milky Way might look from afar.

Half of the spirals have a bar-shaped bulge instead of a spherical or circular-shaped bulge. This is something that evolves over time apparently in some spirals—in some that are more massive it evolves more quickly, in some that are less massive it seems to take somewhat longer—but this is a way in which the stars in the centers of these galaxies, some of them develop these elliptical orbits and they lead to forming this bar-shaped bulge in the center of these galaxies. It's interesting in the fact that the Milky Way itself may actually have such a central stellar bar; indeed, there's good evidence that our galaxy is a barred-spiral galaxy, although, of course, it's very hard to determine that from where we live looking at the center of the Milky Way.

Thinking about ellipticals, spirals, and the irregulars, it turns out the galaxies are not scattered randomly across the sky; Hubble could see this rather quickly as he studied the spiral nebula. They're usually found in clusters ranging from groups of a few to assemblies of thousands. The galaxies nearest our Milky Way form a cluster called the Local Group. The Local Group consists of 40 or so galaxies. It's dominated by three big spirals: the Milky Way, Andromeda, and M33, the Triangulum Galaxy. The rest of the galaxies in the Local Group are dwarf ellipticals, irregulars, and dwarf spheroidals. It's important to understand: Even though the spirals are just 3 of the 40 some galaxies in the Local Group, most of the mass in the Local Group resides in these 3 big galaxies; they dominate the Local Group. The Local Group is actually quite large; it has a radius of about 5 million light-years. So with 40 or so galaxies spread out over some 5 million light-year radius, the space between the galaxies in the Local Group is far vaster and far emptier than the space between the stars in the interstellar medium.

Now let's go to the Sombrero Galaxy. This galaxy lies far beyond the Local Group at a distance about 10 times that of Andromeda. In this sky image, which is about twice the size of the full Moon, the Sombrero seems to stand out as an island universe. That's why I love an image like this; it really hits home that this galaxy really does seem to be an island universe. Most of the points of light you see in this wide-field image are foreground stars; it really hits home that galaxies are island universes. The Sombrero is one of the 20 brightest galaxies on the sky and is listed in the modern version of the Messier Catalogue as M104. The Hubble image of the Sombrero—the HST image of the Sombrero Galaxy—provides a spectacular close-up view of this nearly edge-on spiral, with this pronounced central bulge that goes into the halo of this galaxy and tightly wound spiral arms; and the dust associated with them, you see the dust bisecting the galaxy in a prominent dark lane. This image is a mosaic of 18 images at six different sky positions. The sky width here of this image is about one-third that of the full Moon. The amount of time that HST invested in putting this image together is about 10 hours. The image here is a composite of images taken through red, green, and blue filters, effectively yielding this image in terms of a color that your eye would see if it was bright enough to activate the color receptors in your eye. In terms of physical dimensions, the Sombrero disk here is about 70,000 light-years across, which makes it somewhat smaller than the Milky Way in size, but its mass is a bit more than the Milky Way; the estimated mass of the Sombrero is about 800 billion Suns.

Of course, one of the obvious things about the Sombrero is the glow around it—this big bright glow in the bulge and halo of this galaxy—and it's due literally to billions of unresolved stars even with HST. But HST is capable of resolving out globular clusters in the halo of this galaxy, and there are lots of them: There are 2,000 globular clusters in the Sombrero Galaxy; that's over 10 times more than the Milky Way. There's yet another feature that kind of stands out in this image: If you look towards the nucleus, note how bright that nucleus is; it really stands out as a very bright nucleus. It's interesting because that very bright nucleus harbors a billion solar mass black hole. How do we know this? One of the things that HST's able to do with parts of its very high spatial resolution on the sky is take spectra in very tiny regions. By focusing the spectrographs on Hubble to the core of this galaxy and studying how fast the stars are moving at the center of this galaxy, what was determined was that these stars are moving very, very fast; they're rotating very fast around something that has to be very small and very massive. In other words, they're rotating around something that has to have a billion solar masses, but it has to be in a relatively small place, basically in something the size of the solar system. The only thing we know that you can stuff a billion solar masses into something the size of a solar system is a black hole; and a black hole is essentially an object that's so massive and so compact that at its effective surface, the escape velocity from the object is greater than at the speed of light. In other words, the black hole defines a region where the escape velocity from its gravity is so large, not even light can escape; that's why it's called a black hole.

How such super-massive black holes form at the center of galaxies including the Sombrero is not well understood; but one of the key things we have noticed from studying galaxies with Hubble and other ground-based telescopes is such super-massive black holes are common at the centers of large galaxies. It turns out the Milky Way also has a massive black hole at the center of our galaxy. It's not as massive—it has a mass on the order of 4 million Suns; we've determined that by studying the stars near the center of our galaxy and infrared wavelengths—but one of the other key ways one can find a black hole at the center of a galaxy, one of these super-massive black holes, the black hole itself doesn't emit radiation, but their massive gravity attracts gas at the centers of some galaxies at massive accretion disks, and these massive accretion disks as gas spirals in it heats up the temperatures of millions of degrees and we see phenomena such as x-ray and radio emission from the centers of galaxies that have them, like the Sombrero Galaxy.

Within 100 million light-years of the Milky Way, there are thousands of large galaxies similar in size to the Sombrero, and tens of thousands of smaller galaxies. Some of the large galaxies are isolated or field galaxies in space; some of them dominate small groups like the Sombrero; while others are in rich clusters of galaxies. The Sombrero itself is actually located halfway between the Milky Way and the Virgo Cluster of galaxies. The Virgo Cluster is the nearest rich cluster of galaxies. It contains about 150 large galaxies and has a radius of about 7 million light-years. In other words, it has a lot more galaxies than the Local Group, and is just somewhat larger in terms of size. The distance of the Virgo Cluster is 60 million light-years from Earth. Let's put that big number into perspective; let's think about it in the context, let's imagine there's someone on a planet around a star in one of these galaxies of the Virgo Cluster and they have a really big telescope and they're pointing it at Earth right now. What would you see if you were at the Virgo Cluster right now and you had a big enough telescope to look at Earth; what would you see? You'd be seeing Earth as it was 60 million years ago; in other words, you'd be seeing the Earth certainly after the asteroid impact that wiped out the dinosaurs and 60 percent of the life forms on Earth. That's what anybody living in the Virgo Cluster looking at the Earth right now is seeing; once again, to put these kinds of distances into perspective.

What exists between these groups and clusters of galaxies? Essentially nothing; voids of nearly empty space spanning millions of light-years. The average density of this extragalatic space is less than one atom per cubic meter; much, much less dense than interstellar space and even less dense than space in between the galaxies in our Local Group. Imagine that the Earth was located halfway between the Milky Way and the Sombrero Galaxy. What would the night sky look like to the naked eye? It would be dark;

you wouldn't see any stars at all. The sky would be essentially complete darkness. Only with a telescope at such a location could we see even the nearest galaxies. It's this perspective, thinking like that—putting the Earth out to a place like this—that you truly get the perceptive that galaxies are like islands in a vast cosmic ocean of emptiness.

In summary, it took a 100-inch telescope and help from Henrietta Leavitt and others for Edwin Hubble to chart the nearest galaxies in the 1920s and begin to discover how vast the universe truly is. Today, utilizing the space telescope that bears his name, astronomers are studying nearby galaxies like the Sombrero in unprecedented detail and charting ever more distant galaxies. By studying these galaxies in detail, we're learning things about the super-massive black holes at their centers and studying things like star formation in distant galaxies. This is tremendously important work, and it's also important to consider what even more distant galaxies look like. The motivation for this latter task dates back to Edwin Hubble's even more astonishing discovery in 1929 that our universe of galaxies is expanding.

Next time, we'll revisit this discovery of Hubble with a voyage to galaxies near and far with HST and explore its amazing revelation that the universe is expanding faster today than it was billions of years ago. Please join us then.

Hubble's View of Galaxies Near and Far
Lecture 8

The galaxies that are farther away are moving away faster. The way we can understand this in terms of our universe and an expanding universe is that the larger the distance of the galaxy from us, it takes longer for the photon to get here, and the longer it takes for the photon to get here, the longer the photon is traveling through this expanding space.

A fter Edwin Hubble discovered the vast cosmos of galaxies beyond the Milky Way, he began investigating the velocities of the galaxies. Initial velocity measurements of a few galaxies by others indicated that most were moving away from the Milky Way, some at high speed but with no clear pattern. In his most astonishing discovery, Hubble subsequently showed that there was a simple pattern, and it was tied directly to the galaxy distances in the form of an expanding universe.

Hubble's greatest discovery built upon the work of Vesto Slipher, who used the spectrograph to make the first serious study of galactic velocities in 1912 at Lowell Observatory in Flagstaff, Arizona. A spectrograph takes incoming starlight, breaks it up into a spectrum of colors, then records the brightness of this starlight in each color in wavelengths. These wavelengths evidence the **Doppler effect**: If a star is moving away from you, the wavelengths are redder, or redshifted; if a star is moving toward you, the wavelengths are blueshifted. After obtaining spectra for 14 nebulae, Slipher noted that all the nebulae

© iStockphoto/Thinkstock.

Edwin Hubble's definitive finding of a vast cosmos of galaxies beyond the Milky Way was an achievement that would be the pinnacle of any astronomer's career.

exhibited redshifts and all seemed to be moving away. This observation seemed to favor the idea that the nebulae were not part of this galaxy, but lacking the distances to those nebulae, nothing certain could be said.

Later, Hubble published his work on the Andromeda Galaxy and settled the question of where the spiral nebulae were once and for all; then, Hubble followed up this work by measuring the distances to other nearby galaxies whose velocities had been determined by Slipher. In 1929, he published a landmark paper showing that there was actually a simple linear relationship between the velocities and distances of 24 nearby galaxies. According to this relationship, if the velocity is doubled, the distance is doubled.

Hubble, with his access to the 100-inch telescope at Mount Wilson, began measuring the distances and velocities of galaxies beyond Slipher's nearby sample. By 1936, he and his assistant, Milton Humason, had pushed this relationship between velocity and distance out to velocities of 20,000 kilometers per second, and they found that it held true for distances beyond 100 milllion light-years. There was now no doubt that this was a large-scale effect, and the relationship between the velocities and distances of galaxies became known as **Hubble's law**.

With Hubble's law and the observation that all the galaxies appear to be moving away from the Milky Way, the simple explanation is that we live in an expanding universe.

With Hubble's law and the observation that all the galaxies appear to be moving away from the Milky Way, the simple explanation is that we live in an expanding universe. The most challenging aspect of refining Hubble's law and extending its reach has been the accurate determination of galaxy distances further and further away. The value of **Hubble's constant**—the constant of proportionality between the velocity of a galaxy and its distance—was uncertain for many years after Hubble's initial work. Determining this constant was made a key project of the HST, and by 2009, it had succeeded to a significant degree, finding that Hubble's constant is equal to 74.2 +/– 3.6 kilometers per second per megaparsec.

Images of such galaxies as NGC 3370 were key to enabling Hubble to map the expansion of the universe with unprecedented precision out to vast distances. Even at the great distance of NGC 3370, the Hubble could resolve the Cepheids and, using them, determine the distance of NGC 3370 to be 98 million light-years. This is far beyond the capabilities of ground-based telescopes, which are typically limited to 20 to 30 million light-years.

There's another reason this galaxy is so important: A Type Ia supernova was seen in NGC 3370 in 1994. This kind of supernova occurs when a white dwarf in a close binary star system explodes as a result of taking on additional mass from the larger star. At maximum brightness, these stars are the most luminous kind of supernovae, with the potential of being detected at distances of billions of light-years; also at maximum brightness, they seem to have the same luminosity. We know this because the galaxies in which some of these supernovae have occurred contain Cepheids with well-measured distances. As a result, Type Ia supernovae have become a means to measure very great astronomical distances.

This development has been vital to establishing an accurate value for Hubble's constant, and using this accurate value, Hubble has shown that the cosmos is actually expanding faster now than it was billions of years ago. This finding stunned astronomers in the 1990s; if anything, they had expected the universe's expansion to be slowing down over time. The source of this observed acceleration in the expansion of the universe is completely unknown, although astronomers believe that energy associated with the fabric of space itself is driving the acceleration—the mysterious **dark energy**. ■

Important Terms

dark energy: The mysterious energy believed to be driving the observed acceleration in the expansion of the universe.

Doppler effect: The wavelength shift in the spectrum of a light source as that source moves toward or away from an observer.

Hubble's constant: The constant of proportionality between the redshift velocities of galaxies and their distances.

Hubble's law: The linear relationship between the redshift velocities of galaxies and their distances that is indicative of an expanding universe.

Suggested Reading

Bartusiak, *The Day We Found the Universe.*

Christianson, *Edwin Hubble.*

Kirshner, *The Extravagant Universe.*

Questions to Consider

1. Why are Type Ia supernovae a more reliable standard candle than Type II supernovae?

2. How have HST observations of galaxies both near and far been crucial to the discovery of dark energy?

Hubble's View of Galaxies Near and Far
Lecture 8—Transcript

Welcome back to our voyage of discovery with the images of the Hubble Space Telescope. Last time, we traveled to the Sombrero Galaxy with HST and discussed the identification and characteristics of such star islands in the local universe. Edwin Hubble's definitive finding of a vast cosmos of galaxies beyond the Milky Way was an achievement that would be the pinnacle of any astronomer's career.

But Hubble wasn't done. He tried new approaches to pin down the distances of further and further galaxies that challenged even the Mount Wilson 100-inch telescope in terms of their faintness and tiny angular size. Hubble was also curious about the velocities of the galaxies. Initial velocity measurements of a few galaxies by others indicated that most were moving away from the Milky Way, some at high speed. At the time, it was not obvious that there was any pattern in these velocities. In his most astonishing discovery, Hubble subsequently showed that there was a simple pattern and it was tied directly to the galaxy distances in the form of an expanding universe.

The range of such distances is best appreciated in those HST images that involve deep exposures of sky fields with galaxies near and far. Among these images, the HST view of the nearby spiral NGC 3370 stands out in my mind. Beyond the star-sprinkled arms of this 70,000 light-year wide spiral, we see a more distant edge-on spiral and a face-on barred spiral amidst a host of even more distant galaxies. The image makes my top 10 list not only for its beauty in this perspective, but also because NGC 3370 has played an important role in helping HST to accurately measure the distances to galaxies far beyond the reach of Edwin Hubble. In today's lecture, we revisit Hubble's discovery of an expanding universe and discuss how these HST measurements have revealed a surprising acceleration in the expansion.

Well before the spiral nebulae became known as galaxies, the astronomer Vesto Slipher began the first serious study of their velocities in 1912 at Lowell Observatory in Flagstaff, Arizona. Slipher had earned his Ph.D. in astronomy just a few years earlier from the University of Indiana and he was originally hired at Lowell in 1901 to operate a new spectrograph on the Observatory's 24-inch telescope. Let's review again what a spectrograph does: It takes incoming starlight and breaks it up into its colors. It then records the brightness of this starlight in each color as function essentially of wavelength. How can you use this information to measure the velocity of a star or a galaxy? A key thing to recognize is that stars exhibit spectral features due to their atoms; so stars have absorption lines in their spectra due to atoms in their atmospheres. These absorption lines can shift in wavelength due to a physical principle called the Doppler Effect. All of you have experienced the Doppler Effect. If you're walking down a street and an ambulance goes by, you may have noticed that the pitch of the siren changes; as the ambulance is coming towards you, you hear it at a certain pitch, and then just as it passes it changes its pitch. What's going on is as the ambulance is coming towards you, the soundwaves are getting scrunched, are getting shorter; and then as the ambulance pulls away from you, the soundwaves from the siren are getting stretched, are getting longer, and you hear that as a change in pitch. The Doppler Effect also applies to light: If a star is moving away from you, the wavelengths of light from the star get stretched, they get longer, in other words they go redder; we say that the features get redshifted. If a star is moving toward you, the wavelengths get scrunched, they get shorter; shorter wavelengths mean bluer light, we say the features get blueshifted. It's important to recognize that the amount of this wavelength shift due to the velocity toward you or away from you correspondingly gives you the radial velocity of the star toward you or away from you. This is a fundamental thing that can be measured with spectra: the radial velocity of stars or galaxies.

Slipher at Lowell became an expert in the use of this spectrograph on the 24-inch telescope, and he adapted it for the use on the faint spiral nebulae. This was a lot of work by Slipher just to tinker with the telescope, tinker with the spectrograph, because this was a hard job to do with just a 24-inch telescope, being able to actually collect enough light from these faint nebulae and get spectra out of them; indeed, he would spend over 20 hours getting individual spectra on each spiral nebula. Slipher spent some years doing this work, and he got to a point where he had spectra for the first 14 nebulae; and he stopped and looked at them, he said, "This is interesting, because essentially all the nebulae that I've observed exhibit redshifts; they seem to be all moving away from us for the most part." Among these, the Sombrero Galaxy, M104, had the highest velocity; it was moving at a speed of 1,100 kilometers per second; 1,100 times faster than a rifle bullet away from us. What's more, the average redshift velocity among Slipher's initial sample was 400 kilometers per second; on average, these spiral nebulae that Slipher sampled were moving away from us 400 times faster than a rifle bullet. This velocity was much higher than that of galactic stars.

This was a real puzzle, right? Because the question, of course, at Slipher's time was: Were these spiral nebulae part of the Milky Way or something else? If the spiral nebulae had much, much higher velocities and they were all mostly redshifted compared to the galactic stars that might argue that maybe they don't have anything to do with the Milky Way Galaxy. Slipher presented these results at the 1914 meeting of the American Astronomical Society at Northwestern University. I've included here a slide taken of the audience at this meeting because this particular image I find interesting for a number of reasons: First of all, personally, it's taken in front of the building where my office currently is at Northwestern, in front of the Dearborn Observatory building. It's also interesting from the perspective of the American Astronomical Society, which is the society of all professional astronomers in the U.S. and elsewhere in the world. Currently there are some 7,000 astronomers that are members of the American Astronomical Society; and it's funny to look at this 1914 picture and see here's a picture of the group attending this particular meeting and you can get them all in a picture. Our meeting these days, we meet twice a year and typically 1,000–3,000 astronomers attend these meetings; you can't really imagine taking a picture of that big a group.

What's also fascinating about this image is if you look closely, you see a young Edwin Hubble in the group. He was then a grad student in astronomy at the nearby University of Chicago, and it may well be that by attending this meeting and listening to what Slipher had to say, these results that Slipher presented may well have helped to stimulate Hubble's interest in the spiral nebulae. Indeed, Slipher's results were so amazing, he received a standing ovation after he gave his talk; and Slipher and others interpreted his results to strongly favor the island universe hypothesis for the spiral nebulae. But nothing really definitive could be claimed at that time, because without the distances, you couldn't really say for sure.

Of course, as time went on, Hubble originally did publish his work on the Andromeda Galaxy and settled the question once and for all where the spiral nebulae were; but then what Hubble did to follow up on this work was he began to measure the distances to other nearby galaxies whose velocities had been determined by Slipher. In 1929, he published a landmark paper showing that there was actually a simple relationship between the velocities and distances of 24 nearby galaxies. Let's look at the key plot from Hubble's paper; note what we're plotting here is velocity on the y axis and distance on the x axis, and you may note the units on the x axis for distances are parsecs. A parsec is another unit of distance that astronomers typically use for the big distances in astronomy, and it's typically used a lot in extragalactic astronomy because it's somewhat bigger than a light-year; indeed, 1 parsec is equal to 3.26 light-years. As you look at this diagram, the data is actually the dots, and you see there's actually a lot of scatter in this diagram; and that kind of reflects how hard it was to measure the velocities and the distances back in Hubble's time. But nevertheless, if you fit a line to this data, you see that there is a linear relationship; the line doesn't do a bad job of fitting the data. With this linear relationship, what it's essentially telling you it's something like you double the velocity, you double the distance. This was an amazing result, this relationship between velocity and distance, but this was just for nearby galaxies. The obvious question is how far out did this relationship hold true?

At this point, Slipher couldn't do much more; he was at his limit with the 24-inch telescope at Lowell, he couldn't go toward any fainter nebulae at greater distances. Essentially it was up to Hubble, with his access to the 100-inch telescope at Mount Wilson, to really push out further; to not only measure the distances but also the velocities of galaxies beyond this nearby sample of 24. In this effort, though, he was aided by Milton Humason. Milton Humason originally came to Mount Wilson, he did handiwork originally with the Observatory leading mule teams to bring stuff up to the Observatory, then he was a janitor, then became a night assistant, and eventually became a staff astronomer at Mount Wilson. Humason was an expert at taking long photographic exposures; long exposures to the spectra of these very faint nebulae. Together with Hubble and Humason, by 1936, they had pushed this relationship between velocity and distance out to velocities of 20,000 kilometers per second; and they found that this relationship held true for distances beyond a 100 million light-years. This relationship between the velocity and distances of galaxies, which then became known as "Hubble's Law," there was now no doubt that this was a large-scale effect.

But what did Hubble's Law mean? Why are all the galaxies racing away from the Milky Way? Just the face of it, it makes you think there's something special about the Milky Way; that all the other galaxies seem to be moving away from us. Humanity's gotten in trouble before when it thinks we're something special—thinking that the Earth is at the center of the universe, thinking that the Sun is at the center of the Milky Way—and observations and some thinking later found that no, that's not really the way the universe works. Indeed, given these results, this relationship of Hubble's law and this observation that all the galaxies appear to be moving away from it, as people began to think about this it was realized there's a simple explanation for this, and the simple explanation is we live in an expanding universe. The best way to think about Hubble's Law in the context of an expanding universe is imagining the universe as an expanding balloon and the galaxies are all painted as dots on the balloon's surface. If you imagine you're at one of these dots as the balloon expands, from your perspective on one of the dots, all the other dots will appear to be moving away from you. In other words, this observation that all the galaxies are moving away from the Milky Way, you would get the same observation no matter what galaxy you called home.

Also, there's the other relationship—the other idea here—that the galaxies that are further away will are moving away faster. The way we can understand this in terms of our universe and an expanding universe is that the larger the distance of the galaxy from us, it takes longer for the photon to get here; and the longer it takes for the photon to get here, the longer the photon is traveling through this expanding space. As it travels through this expanding space, the wavelengths get stretched; and if I stretch the wavelengths, I'm making the light redder, I'm making the redshift. That explains the redshift velocity measured on Earth; the higher redshift velocities is simply the result of the photons taking longer to travel through this expanding space from the more distant galaxies. As we'll talk about in a few lectures from now, this observed expansion of the universe is one of the best pieces of evidence that the universe began with a Big Bang some 13.7 billion years ago.

In the years since Hubble's groundbreaking discoveries, the most challenging aspect of refining Hubble's Law and extending its reach has been the accurate determination of galaxy distances further and further away. The velocities are easier to measure in the context of—even for very distant, faint galaxies—if you have a large enough telescope with sensitive enough instruments, you can measure those velocities; it's a technological challenge to measure the velocities. But to measure the distances, that's harder; you have to make some assumptions there, and it's a really hard thing to get the distances to further and further galaxies.

The basic approach to measuring the distance to a galaxy is to identify some object within the galaxy whose luminosity can be assumed and then determine its distance from the object's measured brightness. How do we do this? What Hubble did, for the nearest galaxies he used the Cepheid period-luminosity relationship. What he did is he looked at nearby galaxies, and he'd look for stars that varied in brightness; and he'd measure the period of their variations and he'd measure their brightness. Then he'd make an assumption: He'd assume that the Cepheid period-luminosity relationship that applied to the particular galaxy he was measuring the Cepheids was the same as the Cepheid period-luminosity relationship that Henrietta Leavitt found in the small Magellanic cloud and that was established in the Milky Way Galaxy. What he did: He has the period from this Cepheid in a particular galaxy; he uses the Cepheid period-luminosity relation; he says, "Aha, now I have the luminosity for that Cepheid, I also have the brightness for the Cepheid, which I've measured, and I can calculate the distance to the Cepheid and to the galaxy that the Cepheid is in." This is great, it works for nearby galaxies; but as you go further and further out, eventually with the 100-inch telescope Hubble was working with, the Cepheids would get too faint for him to pull out of the glow of these more distinct galaxies. How could he get the distances to those galaxies? In order to get the distance, he needed something more luminous than a Cepheid variable. Hubble thought about it and he said, "I know what I'll do; I'll use the brightest star in a galaxy." Hubble assumed that the brightest star in any given galaxy has the same luminosity from galaxy to galaxy to galaxy.

But first, if Hubble's going to use the brightest star in the galaxy, he has to calibrate this standard candle. The way he does it is he looks at the nearby galaxies whose distances he's already determined from the Cepheid period-luminosity relationship; then he finds the brightest star in those nearby galaxies and he measures its brightness; and then since he knows the distances from the Cepheids, he can calculate what the luminosity of that brightest star is. Then all those galaxies nearby whose distance he knows, he measures, he gets kind of an average; what is the average luminosity of the brightest star in all these nearby galaxies. Then he makes an assumption: Aha, this luminosity for the brightest star in a galaxy will apply not just for the nearby galaxies, but for the more distant galaxies. Then what does he goes out further and just takes the brightest star in a galaxy and he gets the distances for those more distant galaxies. But eventually, he'll get to a distance where he won't even be able to see and identify the brightest star in a given galaxy.

The point you should get out of this is what Hubble did is a brilliant approach; it's a bootstrap approach to getting yourself out to distances further and further out in the universe. Each step, the way you do it is when you reach the limit of one standard candle—it's not luminous to go any further with your given telescope—what you do is identify a new standard candle, something that's more luminous that would get you out further, and then you calibrate it in the nearer galaxies with the known distances. Given that, then what you do is you apply it in farther galaxies as far as you can see, and then that, of course, will run out and you identify another luminous candle to go another step. In essence, in theory, this is a great idea; this is a way to get the distances further and further and further way and extend Hubble's Law tremendously far out into the universe. But there's a problem: Each time you come up and take a step here with a particular standard candle, the assumption you make is an error; there's no such thing as an absolutely perfect standard candle, so there's an error associated with it. Each time you make a step, the errors compound. As you take new candles to go further and further, the errors in your distances get bigger and bigger and bigger. The point is: As a result of this basic bootstrap approach with various standard candles of increasing luminosity, your distances to most distant galaxies get less and less accurate. Also there's another key point to consider: If you look at galaxies that are further and further and further away, the galaxy images themselves—the angular size in the sky—get smaller and smaller and smaller; and by the time your galaxy is just a faint little smudge, how do you identify a reliable luminous candle in that faint smudge with your particular telescope? These were real problems in pushing Hubble's Law out further into the universe.

Due to this problem and the systematic errors associated with the application of various standard candles, the value of Hubble's Constant in Hubble's Law was rather uncertain for many years after Hubble's initial work; but Hubble's Constant (H) is the constant of proportionality between the velocity of a galaxy and its distance. If an accurate value for Hubble's Constant (H) could be determined, the distance to a galaxy could simply be calculated from its measured redshift velocity. Thus as Hubble developed this work through the 1930s and 40s and into the 50s, a lot of astronomers really got interested in it and wanted to follow up on Hubble's work and refine it and push it out deep into the universe. In particular, during the 1950s, two of Hubble's fellow astronomers at Mount Wilson, Walter Baade and Allan Sandage, were indeed doing this kind of work; and in so doing, they found that Hubble, in his efforts, made a couple of errors; not really the fault of Hubble, it's just that working at the limits when you do astronomy you're bound to have problems. In the case of Baade, he found that Hubble had made an error in his Cepheid calibration; there are actually two types of Cepheids and Hubble had made a mismatch. In addition, Sandage found an error in the brightest star approach; indeed, objects that Hubble thought were single stars at great distance with the 100-inch when observed at greater sky resolution with the soon-to-follow Palomar 200-inch telescope, were actually multiple stars, not the single star. Both of these kinds of errors underestimated the distances and effectively overestimated Hubble's constant.

As a result, Hubble's constant fell from a value of 558 kilometers per second per megaparsec during Hubble's time to 75 kilometers per second per megaparsec by 1956. What do I mean by this number? First recognize that a megaparsec is a million parsecs, or 3.26 million light-years, and that a Hubble constant of 75 kilometers per second per megaparsec; what that means is as we look at galaxies, their redshift velocities go up 75 kilometers per second for every increase of one megaparsec in distance. What happened after it was established at 75 in the mid-1950s? Now a lot of astronomers got into the game, and there were all kinds of ideas about new standard candles and applying them out further, and a whole series of observations were made with new, big telescopes, new instrumentation. But the amazing thing is despite the effort of all these astronomers and all this new equipment and tremendous amounts of observing time, between 1956 and the early 1990s there was no resolution of the Hubble constant issue; different teams found values between 50 and 100 kilometers per second per megaparsec. It was not possible to pin it down; it's just that the systematic errors were just too difficult to sort out. As a result of the conflicts from the ground-based observations, determining the Hubble constant was made a key project of HST; and indeed by 2009, HST succeeded enormously in this effort, finding that the Hubble constant is equal to 74.2 plus or minus 3.6 kilometers per second per megaparsec.

How did HST resolve the conflict in determining the Hubble constant? It turns out that galaxies like NGC 3370 were key in allowing HST to map the expansion of the universe with unprecedented precision out to vast distances. The HST image of NGC 3370 is a net exposure including about a dozen images taken over a month to identify the Cepheid variables in the spiral galaxy. Individually, this image you're seeing of 3370 is the sum of these 12; they were taken multiple times over a month or so to identify the Cepheids in this galaxy. The sky width and total of the image is about one-ninth the diameter of the full Moon. It's a composite of images taken through blue, yellow, and near-infrared filters with HST, and in so doing summing them together yields a composite that once again approximates the natural color one would see with your eye if it was bright enough to activate the color receptors in your eye. The total amount of time put into this image with HST if 25 hours. The key thing to recognize here is this is a very distant galaxy; but even at its great distance, Hubble can resolve out the Cepheids. Using the Cepheids, HST has determined that its distance is 98 million light-years. This is far beyond the capabilities of ground-based telescopes. Only HST can see Cepheids out to 100 million light-years; the typical limit of ground-based telescopes even today is more like 20–30 million light-years.

There's another reason this galaxy is so important: A Type Ia supernova was seen in NGC 3370 in 1994. At that time, it was measured very well with ground-based telescopes. What do I mean by a Type Ia supernova? These types of supernova are different than the Type II supernova we talked about earlier when massive stars die and their iron cores collapse.

Type Ia supernovae occur when massive white dwarfs explode; and this only happens to white dwarfs in close binary star systems. A white dwarf is balanced between electron pressure and gravity; electron pressure holds up gravity. But that'll only work up to a certain mass, and that mass is about 1.4 solar masses. At that mass, gravity can overcome the electron pressure and cause the star to begin to collapse. If you have a white dwarf that's just a shade under 1.4 solar masses and you put it in a close binary with a big, bloated red giant star and the interactions between these two stars end up with some of the gas of the bloated red giant falling onto the surface of the white dwarf, the white dwarf can gain enough mass to push it past this point and cause gravity to say, "Aha, I finally have the mass to defeat the electron pressure," and it begins to switch to white dwarf. As the white dwarf begins to contract, it heats up to a point where it can ignite the carbon that makes up the white dwarf through fusion into heavier elements. But the result of this process, though, the thermonuclear carbon fusion takes off at such a rate you essentially have a thermonuclear runaway and the white dwarf explodes.

The resulting explosion is very interesting, because at maximum brightness it's actually more luminous than the Type II supernova. That means that these Type Ia have potential to be an excellent standard candle because they are so luminous; they can be detected at distances of billions of light-years. But even more important than their luminosity being so intense is the fact that they seem to all have the same luminosity at maximum brightness; at the point where they peak in brightness, they all seem to have the same luminosity. How do we know this? It's galaxies like NGC 3370 that allowed us to calibrate it, because NGC 3370 and another galaxy I'm showing here, NGC 3021, both imaged with Hubble, are two of the few galaxies within 100 million light-years that have both been well-measured with Cepheids with HST and have had a well-studied Type Ia supernova go off in them. These galaxies, these few galaxies, are absolutely key to calibrating a Type Ia supernova and allowing us to find that that they are an exceptional standard candle and can be used to get distances to more distant galaxies.

As a result of being able to extend the Type Ia supernova to get distances out to very distant galaxies, this is one of the key points to establishing an accurate value for Hubble's Constant. It also gives astronomers an opportunity to determine the extent to which the universe has expanded at a constant rate deep into the past. Why has allowed HST to make such a big difference versus the ground telescopes? The real bottom line here is: With HST observations, we've reduced the whole problem of getting distances to distant galaxies to basically two steps: Cepheid variables and Type Ia supernovas. We can use Cepheids—the Cepheid period-luminosity relation—to get the distances to galaxies out to 100 million light-years. That's a big volume; that's a big enough volume that there are galaxies within that volume that there have been Type Ia supernovae that have gone off over the past 20–30 years that have been well-observed and well-calibrated so that we can establish that they have the same luminosity. Then you can use the Type Ia supernovae to go beyond 100 million light-years out literally to billions of light-years. By reducing the problem to just two steps, we've minimized the multiple step errors that were common before. In addition, both Cepheids and the Type Ia supernova are essentially the two best, most well-calibrated standard candles. Those things together have helped reduce the errors and allow us to get such an accurate value for Hubble's Constant with HST. Note also that HST's sky resolution is absolutely key for the distant supernovae. The most distant galaxies are no more than faint little smudges, even with HST; but HST's fantastic sky resolution can pull out these distant supernovae and better separate, isolate, and measure the supernova brightness from the tiny background galaxy.

These measurements that have gone on for a number of years with HST eventually revealed a big surprise in addition to arriving at such an accurate value of Hubble's Constant. The big surprise is: Remember, Hubble's Constant gives you essentially the rate at which the universe is expanding, and it gives you the rate that the universe is expanding and you can trace it far back into the past as far as you can measure it; and with these observations, HST has shown that the cosmos is actually expanding faster now than it was billions of years ago. This was an absolute surprise; astronomers were completely stunned with this result as it developed in the 1990s. If anything, astronomers expected the universe's expansion to be slowing down over time. The idea in the context of the Big Bang is that the Big Bang happened, and that the expectation was the amount of mass in the universe would slow the expansion over time; the gravity associated with the amount of mass in the universe would slow it. The analogy would be like you throw a baseball up, baseball goes up and it slows down. Of course, on Earth it comes back down, but it slows due to Earth's gravity. That's what we expected the universe to do. Instead, the analogy to what the baseball's doing: You throw the baseball up and it rockets up; it stars accelerating after you throw it up. Who ordered this?

We don't have a clue. The source of this observed acceleration in the universe is completely unknown. There's a guess—a good guess—that space itself is driving the acceleration; that there's energy associated with the fabric of space itself, and as the universe expands there's more space, more energy, and that's driving the acceleration. This energy has been labeled something: It's called "dark energy." It's a good name because it's dark in a sense you can't see it; it's also dark because we don't really have a clue what's going on. The bottom line: This dark energy associated with the accelerating expansion of the universe is the greatest mystery in modern astrophysics. We don't know what's going on, but there's tremendous excitement about it because once we understand it, we could really open up new insight in how the universe and physics works in our universe.

In summary, when I look at the HST image of NGC 3370, I see how far we have come in understanding the universe of galaxies and how much further we really have to go. From this image alone by eye, you can crudely gauge which galaxies are near and which galaxies are far by assuming that they all have the same physical size as NGC 3370 and that their apparent angular size is thus solely a function of distance. Essentially, your eye is using galaxy size as a standard candle. However, a detailed Cepheid and Type Ia supernova analysis of this and other HST images would show that angular size is a limited candle since galaxies have a range of physical sizes. In such a way, HST has now put the quest for Hubble's Constant into the past and opened a new path of discovery into the mystery of dark energy. Next time, we'll voyage to non-expanding regions of the cosmos with HST and witness the collisions of galaxies. Please join us then.

The Antennae Galaxies—A Cosmic Collision
Lecture 9

One of the really amazing findings about our galaxy over the past 30 years is a significant fraction of stars in our galactic halo exhibiting these kinds of streaming motions that indicate they were once part of smaller galaxies, and these small galaxies have been disrupted by the gravity of the Milky Way into these streams. The Milky Way has, over time, captured a number of small galaxies and shredded them.

Only in the space between the clusters of galaxies and the isolated field galaxies between the clusters does the concept of an expanding universe apply. As the galaxies within a cluster orbit their collective center of mass, there will inevitably be close interactions that may result in collisions. Hubble has imaged a number of these cosmic train wrecks. The most spectacular nearby example is that of the Antennae Galaxies, which represent a snapshot in time of an ongoing collision between two spirals that were once similar to the Milky Way. Another nearby example, the Sagittarius Dwarf Elliptical Galaxy, is close enough to the Milky Way to strongly feel its gravitational influence; the tidal forces associated with that intense gravity are pulling this little dwarf galaxy apart, leaving a stream of stars through the halo associated with the galaxy.

A much more dramatic event awaits our galaxy in about 2 billion years when the Andromeda Galaxy begins to interact with the Milky Way. Andromeda is one of the few galaxies measured by Vesto Slipher that actually exhibited blueshifts.

This interaction will be significant, because both Andromeda and the Milky Way have similar large masses—hundreds of billions of stars in each galaxy interacting with one another, plus all the gas and dust in both galaxies interacting.

This interaction will be significant, because both Andromeda and the Milky Way have similar large masses—hundreds of billions of stars in each galaxy interacting with one another, plus all the gas and dust in both galaxies

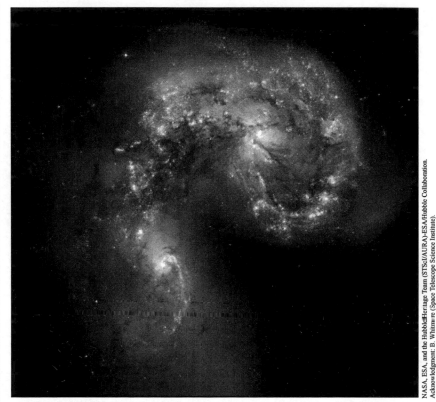

NASA, ESA, and the Hubble Heritage Team (STScI/AURA)-ESA/Hubble Collaboration. Acknowledgment: B. Whitmore (Space Telescope Science Institute).

As galaxies within a cluster orbit their collective center of mass, there will inevitably be close interactions that may result in collisions. The most spectacular nearby example is that of the Antennae Galaxies. These galaxies are a snapshot in time of an ongoing collision between two spirals that were once similar to the Milky Way.

interacting. This is a much more complicated problem to understand than the capture of a small galaxy by a large one.

Such complicated problems are a focus of computational astrophysics, which uses complex calculations to understand the interactions between galaxies. A number of these calculations have focused on the anticipated collision between Andromeda and the Milky Way, and they find that the **elliptical galaxy** that results will have far less gas than the two spirals had originally.

By smashing a number of the gas clouds in the two galaxies together so violently, the collision will ignite the formation of many millions of stars; the result will be a starburst.

How likely is this simulation? A key uncertainty here is the velocity of Andromeda across our line of sight. Remember, the Doppler effect gives us only that component of velocity toward us or away from us. If Andromeda is moving rapidly this way, then it might not hit us at all. The best we have been able to do with estimating Andromeda's movement across our line of sight is to say that its velocity is probably equal to or less than its velocity coming toward us. It's possible, then, that there might not be a head-on collision, just a glancing blow. The collision may not be as dramatic as the simulations have shown. However, even a glancing blow could have a significant effect on the two galaxies.

Computer simulations of galaxy interactions are developed and tested using observations of collisions in progress. As one of the nearest examples of an ongoing collision between two spirals, the Antennae Galaxies provide a wonderful snapshot into the potential future of the Milky Way and Andromeda. A ground-based, wide-field view of the Antennae shows long tidal tails of stars, from which the interacting galaxies earned their name. Seeing these tidal tails on the Antennae Galaxies gives us more confidence in the software simulations; this is something we might expect when Andromeda and the Milky Way interact. Moreover, the high resolution of the HST can identify the individual star clusters that form; indeed, there are large numbers of massive, million–solar-mass star clusters amongst this glow in the collision between these two galaxies.

Hubble's discovery of such interacting galaxies very far away reveals that they were more common in the distant past when the universe was a smaller place. Hubble continues to rapidly expand the detailed image inventory of distant interacting galaxies. One of the most amazing images appears on a poster that shows 59 such interactions between galaxies. We see head-on collisions; we see glancing blows; and we see them in different epochs of their interaction. Some are just starting, and some are well along. These Hubble images are vital to a better understanding of the future interaction between the Milky Way and Andromeda. Although each colliding galaxy photo is in some way unique and only a snapshot in time of a lengthy, dynamic process, the sheer quantity of detailed images provided by Hubble affords astronomers an unparalleled opportunity to test and refine their simulations across the timeline of galaxy interactions. Hubble has made a dramatic difference here. ∎

Important Term

elliptical galaxy: A gas-poor, elliptically shaped system of up to 500 billion typically old stars.

Suggested Reading

Christensen, de Martin, and Shida, *Cosmic Collisions*.

Loeb and Cox, "Our Galaxy's Collision with Andromeda."

Questions to Consider

1. How might we determine whether a nearby star is a native of the Milky Way or an immigrant from a cannibalized dwarf galaxy?

2. Based on the HST image alone, what are the features that lead us to conclude that the Antennae Galaxies are the result of two colliding spirals rather than two colliding ellipticals?

The Antennae Galaxies—A Cosmic Collision
Lecture 9—Transcript

Welcome back to our exploration of the cosmos with the Hubble Space Telescope. Last time, we discussed how Hubble has mapped the expansion of the universe with unprecedented detail out to distances of 10 billion light-years. It turns out that on much smaller scales, the universe is not expanding. For example, as much as we might want to, we can't blame our expanding waistlines over the years on the universe. Also, we don't observe the Earth or the solar system or the Milky Way or even the Local Group to be expanding over time. Indeed, it's only on large scales in the space between the clusters of galaxies and the isolated field galaxies between the clusters where the concept of an expanding universe applies. In the case of a cluster of galaxies, the mutual gravitational attraction keeping the galaxies together is strong enough to overcome the expansion of the universe in its localized region. As the galaxies within a cluster orbit their collective center of mass, there will inevitably be close interactions that may result in collisions.

Hubble has imaged a number of these cosmic train wrecks. The most spectacular nearby example is that of the Antennae Galaxies, which are at a distance of 45 million light-years. These galaxies are a snapshot in time of an ongoing collision between two spirals that were once similar to the Milky Way. The colorful disruption of the once graceful spirals of these two galaxies testifies to an episode of cosmic violence much more far-reaching than that of an individual supernova. Such a collision is not a short-term event where astronomers can make some popcorn and watch it play out over the course of a night with Hubble; indeed, the interaction of the Antennae began over 200 million years ago. As a colorful example of the ultimate in slow motion collisions, this Hubble image of the Antennae makes my top 10 list both for its visual beauty and its importance in understanding a process that impacts the evolution of many galaxies. In today's lecture, we're going to discuss this image in the context of modeling the gravitational interactions of galaxies and predicting the future of our Milky Way.

Although our Local Group contains only 40 or so galaxies in a very large volume, it does show evidence of past and continuing interactions between these galaxies. This finding is particularly true in the immediate vicinity of the Milky Way. Let's start by looking at an image that's a near-infrared picture of the entire sky. In the near-infrared, you can cut through some of the dust in the Milky Way. When we look at this particular image, we see the disk of the Milky Way as shining very brightly in the near-infrared—and this is essentially a measure of star counts when we look in this part of the spectrum—so we see the Milky Way's disk very brightly, and we see some dust still in the vicinity of the disk of the Milky Way. We also see the bulge at the center of the Milky Way—the bulge of bright stars in the near-infrared—and we also see in the halo of our galaxy the small and the large Magellanic clouds. These two satellite galaxies of the Milky Way are very obvious to naked-eye observers in the southern sky.

But if you look very carefully at this image, you see something else: Just below the bulge of our galaxy, you see a faint smudge. This faint smudge of near-infrared light is the Sagittarius dwarf elliptical galaxy. It's called the Sagittarius dwarf because it's in the constellation Sagittarius, just as the Sagittarius Star Cloud we talked about earlier in the course. This dwarf galaxy is about 70,000 light-years away, and it's about 10,000 light-years across. It's much, much less massive than our Milky Way: about 10,000 times less massive. It orbits our Milky Way about once every 750 million years. The important thing to recognize about this little galaxy is that it's close enough to the Milky Way to strongly feel the gravitational influence of the many hundreds of billions of stars collectively in our Milky Way Galaxy; and essentially what's happening is the tidal forces associated with the intense gravity of all these stars in the Milky Way is beginning to pull this little dwarf galaxy apart. What's happening is as this galaxy has been orbiting the Milky Way, it's been slowing down the motion of this dwarf galaxy in its orbit; and in so doing it transfers some of this energy to the stars making up the dwarf galaxy. By giving the stars this extra energy and giving them more velocity, it's effectively elongating the dwarf and also it's literally stripping stars away from this little galaxy. As a result, as this galaxy continues on its orbit around the Milky Way through the halo of our galaxy, it's leaving stars behind, stars are trailing behind, and so we actually see a stream of stars through the halo associated with this galaxy. Eventually, what's going to happen is the Milky Way will effectively just shred it all to bits, and we're actually seeing this in real time now; we're seeing the slow motion process of the Milky Way shredding this little galaxy.

It's possible to simulate this effect on a computer. You put a galaxy like the Milky Way and a small galaxy and you let gravity go to work on a computer and you can produce a simulation video showing what happens over a large interval of cosmic time. In this particular video of this simulation, we see what's happening over two-and-a-half billion years of this interaction. It shows what's happened over the past two billion years, and what's going to happen in the future half a billion years. What we see is the galaxy is slowly being taken into pieces, bit by bit, stars being stripped away, and the Milky Way is converting this little galaxy into a star stream through the halo. One of the really amazing findings about our galaxy over the past 30 years is a significant fraction of stars in our galactic halo exhibiting these kinds of streaming motions that indicate they were once part of smaller galaxies and these small galaxies have been disrupted by the gravity of the Milky Way into these streams. The Milky Way has, over time, captured a number of small galaxies and shredded them. It's quite likely that almost all the stars in the halo of our Milky Way galaxy are the shredded remains of small galaxies swallowed up by the Milky Way long ago; indeed, some of the globular clusters may also have been acquired from small galaxies that were shredded by the Milky Way. It's quite likely that the Local Group once had appreciably more small galaxies billions of years ago before they were all gobbled up— at least a good fraction of them were gobbled up—by the three big spirals in our Local Group of galaxies. A much more dramatic event awaits our galaxy in about 2 billion years when the Andromeda Galaxy begins to interact with the Milky Way.

When Vesto Slipher did his study of the velocities of the nearby spiral nebula, we talked about last time about how he found that almost all of them exhibited redshifts. But Andromeda is one of the few that actually exhibited blueshifts. Slipher found that Andromeda was moving toward Earth at 300 kilometers per second; 300 times faster than a rifle bullet. But note to understand the actual velocity between our galaxy and Andromeda, one has to take into account the actual motion of the Sun orbiting around the Milky Way; and in so doing, one can also get an idea of our own motion, the motion of spaceship Earth, in our galaxy. This is kind of educational for those of you who like to say you never get to go anywhere; let me give you a sense of how fast you're going somewhere. The Earth rotates on its axis at a speed of about 1 kilometer per second; the Earth goes around the Sun at a speed of 30 kilometers per second (30 times faster than a rifle bullet); and the Sun itself orbits the Milky Way at a speed of 200 kilometers per second (200 times faster than a rifle bullet). So spaceship Earth is going places really fast, and you're along for the ride; of course, most of the space in the galaxy is empty so we're not going anywhere really exciting, but we're going nowhere really fast.

But after you take out these motions of the Earth and the Sun going around the center of the Milky Way, one arrives at the velocity of approach between Andromeda and the Milky Way, and that velocity is 120 kilometers per second. We know that Andromeda's at a distance of 2.5 million light-years; so given the relative velocity between the two galaxies and this distance, we can estimate when the gravitational interaction between the two galaxies will begin. The interesting thing is the interaction should actually begin before the Sun becomes a red giant in about 5 billion years. This will be one heck of an interaction; it won't be like the Sagittarius dwarf being shredded by the Milky Way, a very unequal collision so to speak, because both Andromeda and the Milky Way have similar large masses, hundreds of billions of stars in each galaxy interacting with each other plus all the gas and dust in both galaxies interacting with each other. This is a much more complicated problem to understand than the capture of a small galaxy by a big galaxy like the Milky Way.

Such complicated problems—like the interactions of two systems with hundreds of billions of stars and trying to understand the gravity between those two systems plus the hydrodynamics between the gas clouds in such galaxies—are a focus of computational astrophysics, a really key and emerging field in astronomy today. Computational astrophysicists use powerful computers to carry out the complex calculations needed to understand the interactions between two big galaxies like Andromeda and the Milky Way. In many ways, the field of computational astrophysics is technology-driven, just like observational astronomy. Observational astronomers always want to build bigger telescopes and have more sensitive instruments so they can understand more detail about stars and galaxies and nebulae in the universe and study galaxies that are further and further away. Computational astrophysicists want more and more powerful computers, because with more powerful computers they can deal with the more sophisticated software needed to take into account not just gravity but the hydrodynamics associated with gas flows and put all this physics into it. Furthermore, with a more powerful computer you can study galaxies not just in terms of a tiny sample of each galaxy—only a handful or hundreds of stars in each galaxy—but you can actually put in many, many stars and study many more interactions so you can get a much more detailed model of really what's happening with these more powerful computers.

Furthermore, with the more computer power, you can study the interaction at higher time resolution and also study tinier pieces or break each galaxy up into tinier pieces so you can study in much greater detail how each piece of the two galaxies will interact with each other and follow it through with a long set of time steps. This is an area, computational astrophysics, which has really hit its own in terms of studying things like collisions between galaxies, and we can really see a point where the actual models, you can compare them with the observations and really continue with better and better computers and better observations to refine these models more and more and really begin to reproduce the reality we see, particularly with the Hubble Space Telescope.

A number of these models of galaxy interactions have focused on this anticipated collision between Andromeda and the Milky Way; the merger between these two galaxies. Typically what you get out of these two models is the initial encounter between them will yield big tidal tails of stars and gas; as the two galaxies approach one another, the first real signs of it will be you'll see stars and gas in each galaxy be thrown off into extra galactic space, never to return. Then as the galaxies get closer and closer they slow down and they begin a dance over the ensuing some 3 billion years; and then they finally will merge, the Milky Way and Andromeda, into an elliptical galaxy. One can simulate this, of course, on a computer; and here we see a simulation video spanning 15 billion years—from the past 5 billion years to the next 10 billion years—of this one anticipated expectation of the interaction between Andromeda and the Milky Way. Here we break it down both in terms of the gas in both galaxies and what happens in the stars in both galaxies.

What will happen in the result of this—the elliptical galaxy that ends up combining many of the stars in M31 and the Milky Way—it will have far less gas than the two spirals had originally. What happened to the gas? By smashing a number of the gas clouds in the two galaxies together so violently, this collision shock will ignite the formation of many, many, many millions of stars; there will be a starburst. One of the heralds of this collision when the bulk of the two galaxies smash into each other will be all these gas clouds whacking each other, collapsing, and just a burst of all kinds of stars; so a lot of the gas will go into new stars rather quickly as a result of this collision. Some other gas will just be tidily expelled off into interstellar space. In terms of star on star collisions in the merger between the two galaxies, remember there's a lot of empty space between the stars in both galaxies, so there will actually be few star-on-star collisions. The real thing that really jumps out at you when two galaxies like this collide is the starburst associated with the rapid formation of many stars as a result of the collisions of the clouds in the two galaxies.

What will be the fate of the Sun in this interaction? There are a couple of possibilities. One, it could be tidally expelled off into intergalactic space; just thrown away from the two spirals and just go on its own way off deep into intergalactic space. The most likely possibility, though, is that the Sun—and the Earth along with it—will end up in an outskirt orbit of the merged galaxy. Right now, we're already in the suburbs of the Milky Way, 28,000 light-years away from the center; but it's likely that in this new merged elliptical galaxy, the Sun will live much further way, it will be in the exurbs of that galaxy. From that perspective, the night sky would look very different. There would no longer be a Milky Way stretching across the sky; instead, there'd just be in one part of the sky this nebulous glow that would be the center of our new elliptical galaxy that we'd call our new home.

Where do we put this particular interaction, this fate for the Sun, on the list of cosmic things to worry about? As we said, the interaction between the two galaxies won't begin for another 2–3 billion years at the earliest. At that timeframe, the Sun is still a main sequence star; but at that timeframe, the luminosity of the Sun will have increased enough so that temperatures on Earth may have reached a point where complex life such as humans may find it hard to exist. Either way, in about 2–3 billion years, things will definitely get kind of dicey for humans on planet Earth. But between now and then, we talked earlier in the course about the rates of asteroid and comet collisions with the Earth—big ones—and certainly on that timeframe, a couple billion years, we'll get hit a number of times by asteroids comparable in size to the one that caused the extinction of the dinosaurs. Of course, there's always the possibility of a nearby supernova going off in the next 2–3 billion years.

Listing these things as we've gone through this course, you might say, "Wow, this galaxy, this universe; this is a dangerous place to live. You get whacked by asteroids, your Sun fries you, you get thrown out of your galaxy, a supernova goes off nearby; life must be extraordinarily rare in the universe and in the galaxy." Actually, most scientists think the exact opposite; they think that life is probably quite common in the galaxy. How do we make sense of this? If you look at the Earth, it's been around for 4.6 billion years, and life has existed on Earth for at least the past 4 billion years. Life somehow formed on the Earth rather quickly after the Earth came into being, and life on Earth has survived for some 4 billion years, taking the best that the universe has had to throw at it: asteroid hits, comets, everything. The Earth just on its own sometimes gets hotter so it's a sauna everywhere, or it freezes up like an ice cube; we've gone through epochs on Earth when temperature conditions have been like that. Life has survived; life is resilient; life can find a way. That's one of the amazing things when one studies life on Earth. So the viewpoint is as violent as the cosmos appears to be in places, understand clearly the time scales we're talking about, and also understand the resilience of life.

Also, although Earth in terms of humans, complex life is much more fragile, Earth has finally over 4.6 billion years developed a species that is intelligent, and it may well be that as an intelligent species, we can stop the very thing that took out the dinosaurs after the ruled the Earth for 150 million years. Humans have developed the intelligence and soon may have the technology to stop asteroids from hitting us; indeed, stop the top the thing on a list of cosmic things to worry about from happening. That's the good news about this; so don't let all this cosmic violence freak you out. This is just the way the universe is and life has found a way, at least on Earth and probably elsewhere, to deal with it.

Let's get back to this collision between Andromeda and the Milky Way; how likely is this simulation? There's a key uncertainty I haven't talked about yet, and that key uncertainly is the velocity of Andromeda across our line of sight; the component of its velocity across our line of sight. Remember the Doppler Effect only gives us that component of velocity towards us or away from us. If Andromeda's moving really fast this way, then it might not hit us at all. How do we get a measure of how fast Andromeda's moving across our line of sight? They way you'd do it is you'd have to take a picture of Andromeda and then come back at some future time and take another picture of Andromeda and see how much the stars have moved, and given the distance of Andromeda calculate its velocity. That's easy to say, but it's very hard to do, because Andromeda is so far away that in terms even of Hubble Space Telescope measuring tiny angles on the sky, it can't see any change in the motion of M31 across our line of sight over the past some 20 years that Hubble's been in orbit; it's just too far away. Indeed, the best we've been able to do with estimating Andromeda's movement across our line of sight is to say that its velocity is probably equal to or less than its velocity coming toward us. What that means is it's possible that as Andromeda moves toward us it may be moving enough this way so it will be a glancing blow; there might not be a head-on collision, it just may be a glancing blow. The collision may not be as dramatic at all as the simulations have shown. However, even a glancing blow could have a significant effect on the two galaxies.

Also in considering this particular interaction, how can we test the simulation software itself and applying it to something that's going to happen in the future? The way we test and develop this software on computers simulating these interactions with galaxies is apply it to observations to collisions in progress. As one of the nearest examples of an ongoing collision between two spirals, the Antennae Galaxy provides a wonderful snapshot into the potential future of the Milky Way and Andromeda. A ground-based, wide-field view of the Antennae shows us these long tidal tails of stars from which it earned its name. These tails stretch over a half Moon-width on the sky; in terms of real physical dimension, that works out to be two Milky Way widths. These tails originated over 200 million years ago. These kinds of tails, we said before, are common in interactions between galaxies; if you have two spiral galaxies colliding with each other, this is part of their initial interactions, spinning off these tidals of stars and gas into the cosmos.

Explicitly, seeing these tidal tails on the Antennae Galaxies gives us more confidence; this is something we might expect when Andromeda and the Milky Way interact.

The Hubble image of the central part of the Antennae gives us a really colorful high-resolution view of the star formation stimulated by the collision of these two gas-rich spirals. What this image is when we look at the Antennae is a composite of images taken through blue, green, red, and pink filters with Hubble; the pink filters actually isolated on the emission line due to hydrogen, H Alpha, at 656 nanometers. The wonderful colors you see in the Antennae image, they roughly approximate the natural color you would see with your eye if it was bright enough to activate the color receptors in your eye. The total amount of time put into this image is about five hours on Hubble. These colors are important because they help us diagnose what's going on in the Antennae galaxies. If we look at the two yellowish regions, they're basically mostly old stars associated with the cores of the two original galaxies. The blue regions are areas of massive star formation associated with the starbursts as these two galaxies' clouds whacked into each other. The nearby pink regions associated with these blue regions are glowing clouds of hydrogen gas, the gaseous nebula excited by the hot young stars produced in these starbursts. As you see amongst these blue regions and pink regions you see the brown stuff, that's the dust associated with the gas involved in these collisions in the starbursts.

I want to say a little bit more about this collision-induced starburst. As we study these interacting galaxies in the form of the Antennae, we can identify with the great resolution of the Hubble Space Telescope the individual star clusters that form; indeed, there's a large number of massive, million solar mass star clusters amongst this glow in the collision between these two galaxies; there are a thousand of these million solar mass clusters. Most of these star clusters have formed within the past 200 million years; the Antennae is actually forming stars over six times faster since then 200 million years ago. The starburst we observe with the Antennae is completely consistent with what we get from our models of colliding these two galaxies; and indeed if we could follow the Antennae for a few more billion years we would expect it to eventually evolve into a gas-poor elliptical as the gas we currently see so violently interacting, a lot of it ages into stars, a lot of it gets thrown out of the galaxy. What we're left with is essentially just stars in this elliptical and some of these star clusters that have formed actually evolve into globular star clusters.

It's important to realize that the Antennae represent just one snapshot in time of one galaxy collision. Fortunately, with its ability to image distant objects in unprecedented detail, Hubble has revealed a menagerie of interacting galaxies that can serve to challenge and improve the collision models. In particular, let's start with the Mice. The Mice are an interacting pair of spiral galaxies at a distance of 300 million light-years, and they exhibit tidal tails similar to the Antennae Galaxies. Remember, at 300 million light-years, this is about seven times further away than the Antennae. This Hubble image is about three Milky-Way-widths across; I'll give you a sense of scale here. What we note is when we simulate on a computer this interaction between these two spirals, we see that the simulation nicely reproduces this particular snapshot in time of this particular interaction between these two galaxies. What we have going on here: These two spirals actually are interacting orthogonally. We can model this kind of geometry—these different geometries can be modeled on a computer quite easily once you get the software going—and putting these models together, we see that the snapshot of Hubble of the Mice is actually we're seeing the collision about 160 million years after the initial close encounter. Just like with the Antennae, we expect the Mice will eventually coalesce into an elliptical galaxy. In another video here, I've taken another model of the Mice interactions—this is a very detailed model—and here we've broken up the different components of the galaxy so you can see how different components of the galaxy comprising the Mice have evolved and will evolve over cosmic time through this interaction. We can look at the old stars and see what happens to them; we can isolate on the gas and see how shocks form in the gas and how we can see where stars might begin to form; and we can also just isolate on the new stars created as a result of this interaction and see what happens to them.

Of course, not all collisions involve similarly-sized galaxies; we've talked about so far in terms of modeling a lot of these interactions between roughly similar-sized spirals. There are other kinds of interactions: The Hubble image of the 420 million light-year distant Tadpole Galaxy shows the damage that even a moderate-sized galaxy can inflict on a big spiral. Here was this spiral galaxy minding its own business and some moderate-sized galaxy comes into it and whacks it. What this galaxy did was it warped the large spiral's disk; and furthermore, the key signature in this image is it has left a 300,000 light-year trail of stars in its wake. This big guy took a hit, he's bleeding, but he's still hanging in there. This exposure's a real wonderful exposure with Hubble; it's over eight hours long and it shows many details. If you look at it, you see these blue knots of young star clusters in the disk of the galaxy and trailing off into its long, long tail. But if you look closely at the other galaxies in the background here, you see something amazing: You see interactions among those background galaxies, too. So what this is telling you, at least in some parts of the universe, such interactions are certainly not outrageously rare.

Although colliding galaxies are rare in our local universe—it does appear that they're relatively rare in our local universe—the serendipitous discovery of such interacting galaxies in the faraway backgrounds of deep HST images like the Tadpole shows that they were more common in the distant past when the universe was a smaller place. We have other evidence: Independently, the high fraction of ellipticals in the inner regions of rich clusters of galaxies is also a likely signature of more frequent collisions long ago. In other words, Hubble has helped us establish a lot of evidence that collisions between galaxies were more common in the past; and Hubble continues to rapidly expand the detailed image inventory of distant interacting galaxies. In doing so, what we've realized is that there are a whole variety of collision geometries. We see some of these collisions edge-on; we see some face-on views; and then we actually see different characteristics among the collisions in progress: We see head-on collisions; we see glancing blows; and we see them in different epochs of their interaction. Some are just starting, and some are well along.

To me, one of the most amazing images put out from the press folks at the Hubble Space Telescope is a poster showing 59 such interactions between galaxies; this is an amazing image. Let's put this poster into perspective: Each one of the collisions we're seeing here typically involve Milky-Way-sized galaxies; if they're a Milky-Way-sized galaxy, each one involves hundreds of billions of stars, and many of them may have planets, and there are gas clouds, other things. As these interactions are happening, millions of new stars are being born; other older stars are thrown off into deep space, never to return from these galaxies. Each one of these interactions is its own drama taking billions of years to play out, in every case leaving local habitats of these interacting galaxies irrevocably changed. Indeed, billions of years from now, the Milky Way and Andromeda may well be a postage-stamp-sized photo in such a collage of interacting galaxies imaged by an alien telescope in a faraway star system.

In summary, HST images like that of the Antennae are key to a better understanding of the future interaction between the Milky Way and Andromeda. Although each colliding galaxy photo is in some way unique and only a snapshot in time of a lengthy dynamical process, the sheer quantity of detailed images provided by HST affords astronomers an unparalleled opportunity to test and refine their simulations across the timeline of galaxy interactions. You really need to understand this; Hubble has really made a dramatic difference here. with its fantastic sky resolution on the sky, it can study even these interacting galaxies where they're more frequent at great distance and get all the kinds of details and all these individual snapshots—different geometries, different epochs in time—and this is fodder for a computational astrophysicist to plug them into their models and refine the models such that eventually we'll be able to model the anticipated interaction between Andromeda and the Milky Way so much better. This work is also essential in that these kinds of collisions now appear to play an important role in the evolution of many galaxies, particularly long ago.

Next time, we're going to voyage with Hubble even deeper into the distant universe utilizing gravitational lenses that are so massive they distort the fabric of space itself. Please join us then.

Abell 2218—A Massive Gravitational Lens
Lecture 10

Hubble also has the remarkable achievement that it's actually imaged the double Einstein ring, which is an alignment of not just two but three galaxies along the sightline.

One of the most fascinating phenomena observable in the night sky with a telescope is a **gravitational lens**. Such a lens is actually an object, or cluster of objects, whose mass is large enough to curve the surrounding space to such a degree that the images of background objects are affected. The idea of curved space sounds like something right out of science fiction, but it is real and its effects can be observed in stunning detail with Hubble.

As part of his general theory of relativity, Albert Einstein developed the idea that gravity is a manifestation of the curvature of space in the vicinity of massive objects. If light is coming toward us in the direction of the Sun from a star beyond the Sun, it will be curved by the curvature of space associated with the mass of the Sun. In 1916, Einstein predicted that starlight right on the edge of the Sun should appear shifted 1.7 arc seconds from its true sky position. During an eclipse in 1919, British astronomers confirmed this prediction.

The simplest case of a gravitational lens occurs when the observer, the lens, and the background object are all perfectly aligned. In this case, "perfectly aligned," means within much less than 1 arc second. The result of this alignment is an **Einstein ring**. The space curvature around a gravitational lens spreads the image of the background object into a ring of light or into arcs as partial rings. About 50 partial to full Einstein rings have actually been discovered optically. The high-resolution images obtained with Hubble are absolutely key to understanding the physics associated with these Einstein rings.

The most spectacular cases of gravitational lensing involve distant rich clusters of galaxies. The Hubble image of the rich galaxy cluster Abell 2218,

The Hubble image of the massive cluster of galaxies Abell 2218 shows a number of thin, arc-like features that seem to partially encircle the large elliptical at the cluster core. It turns out that the arcs are galaxies far beyond Abell 2218, whose images have been amplified and distorted by the intervening space curvature associated with this cluster's gravitational field.

NASA, Andrew Fruchter, and the ERO Team [Sylvia Baggett (STScI), Richard Hook (ST-ECF), Zoltan Levay] (STScI).

2 billion light-years away, reveals about 80 bright cluster galaxies and more than 100 gravitationally lensed, arc-like images of the background galaxies. When we study some of the arcs in detail, we note that some of the faintest arcs are very red, indicating that they are likely to be a great distance at high redshift. Indeed, in 2002, it was possible to obtain brown-based spectra of two of the faint red spots in the Hubble image, and they yielded a very high redshift. The galaxy responsible for these two red spots was 13 billion light-years away and actually much smaller than the Milky Way. The lensing provided by Abell 2218 brightened it by a factor of 30. Without the brightness boost by the gravitational lensing and the sharp eye of the HST to see these thin arcs, these galaxies would have been almost impossible to detect.

Dark matter is matter that interacts with visible matter through gravity but not through electromagnetic forces—not through emitting or absorbing photons. We can't "see" it, but we can sense that it's present through such effects as gravitational lensing.

The lensing mass calculated for Abell 2218 is about 10 times greater than the mass in the stars and the gas, and that holds true not just for Abell 2218 but for other clusters that have been studied through gravitational lensing. What this tells us is that most of the mass in Abell 2218 must be **dark matter**. Dark matter is matter that interacts with visible matter through gravity but not through electromagnetic forces—not through emitting or absorbing photons. We can't "see" it, but we can sense that it's present through such effects as gravitational lensing. Hubble observations of gravitational lenses in other galaxy clusters continue to provide new clues about the character of dark matter.

It has been known since the 1970s that the dark matter is likely to be common in **spiral galaxies**, based on the rotational velocities of their outer stars. The original expectation was that toward the edge of a spiral galaxy, there should be less mass because we see fewer stars and the velocity of those stars should be less, just as the outer planets in the solar system orbit more slowly than the inner ones. But we actually observe a constant velocity or a slow increase out to distances where we don't even see stars in the galaxy. These "flat rotation curves" are best understood as resulting from a dark matter halo that extends far beyond the optical galactic disk and has much more mass than the visible matter in the disk. ■

Important Terms

dark matter: The dominant, unknown constituent of matter in the universe that interacts with visible matter gravitationally but not through electromagnetic forces.

Einstein ring: The simplest case of a gravitational lens, in which the observer, lens, and background object are perfectly aligned.

gravitational lens: An object (or cluster of objects) whose mass is large enough to curve the surrounding space to a degree at which distortions are produced in the images of background objects.

spiral galaxy: A gas-rich, disk-shaped system of young and old stars with ongoing star formation in its characteristic spiral arms.

Suggested Reading

Freeman and McNamara, *In Search of Dark Matter*.

Gates, *Einstein's Telescope*.

Questions to Consider

1. How might we determine whether an odd-shaped galaxy that appears on the sky within a rich galaxy cluster is a gravitationally lensed image of a much more distant galaxy?

2. If the Milky Way has a massive halo of dark matter, why doesn't HST see gravitationally lensed images of galaxies all over the sky?

Abell 2218—A Massive Gravitational Lens
Lecture 10—Transcript

Welcome back to our journey through the universe with the Hubble Space Telescope. Last time, we explored the future interaction of the Andromeda Galaxy with our Milky Way through high resolution Hubble images of the Antennae Galaxies and other star systems currently undergoing the cosmic equivalent of hostile takeovers, mergers, and acquisitions. As in the cases of the core collapse of a Type II supernova and the impact of Comet Shoemaker-Levy 9 with Jupiter, gravity plays a key role in these galaxy collisions. It's also responsible for one of the most fascinating phenomena observable in the night sky with a telescope: a gravitational lens. Such a lens is actually an object, or cluster of objects, whose mass is large enough to curve the surrounding space to a degree where the images of background objects are affected. The idea of curved space sounds like something right out of science fiction, but it is real and its effects can be observed in stunning detail with Hubble.

For many years, the Hubble image of the massive cluster of galaxies Abell 2218 has been the textbook example of a gravitational lens. Among the galaxies in this rich, compact cluster, the image shows a number of thin, arc-like features that seem to partially encircle the large elliptical at the cluster core. Indeed, the pattern of these arcs gives the appearance that the Abell 2218 galaxies are acting in concert like a giant lens. It turns out that the arcs are galaxies far beyond Abell 2218 whose images have been amplified and distorted by the intervening space curvature associated with this cluster's gravitational field. This image is one of those HST vistas, like the Eagle Nebula, where my jaw simply dropped upon the first view. Nothing like Abell 2218 had been seen before because only Hubble could resolve out the faint filamentary images of the lensed galaxies. In today's lecture, we utilize the Hubble image of Abell 2218 to explore the nature of gravitational lenses and their implications regarding dark matter and the distant universe.

At about the same time that Henrietta Leavitt and Vesto Slipher were paving the way for Edwin Hubble to change the astronomers' universe, Albert Einstein was rewriting the physical laws upon which our understanding of the cosmos is based. In particular, Einstein developed the idea that gravity is a manifestation of the curvature of space in the vicinity of massive objects. He did this as part of his general theory of relativity. Let's think about this idea: If you're like me, it's very hard to think in terms of what GR is basically about, because it treats space itself as a four-dimension spacetime, and it's really hard for me to think in four dimensions; and certainly based on everyday life, it's hard to imagine space in terms of four dimensions. Let's put this, the way general relativity treats space, in terms of a three-dimensional example; on discussing this, of course, you're going to be seeing this on a two-dimensional TV screen, so bear with me.

Imagine you have a flat, flexible sheet of rubber for as far as the eye can see. Imagine further that I took a baseball and I put a baseball on this sheet. What you would notice is the sheet would no longer be flat; it would be curved underneath the baseball. Imagine I took a marble and put a marble on this sheet with the baseball there. What would happen to the marble? It would roll into the baseball. It feels curvature in its vicinity; in a sense, space curvature. Imagine I took the baseball off the sheet and I just put the marble down. It wouldn't roll, it would just stay there; but once I put the baseball down it feels the curved space in the vicinity of this object with mass and rolls into it. Likewise, if I imagine I put this marble on the sheet and I gave a little push with the baseball on the sheet, the marble would feel curvature and its direction of motion would change. Depending on its velocity, it might actually roll right into the baseball. On the other hand, if I took the baseball off the sheet and I gave that marble a little push, it would just keep rolling down the sheet; and indeed, if there was no friction between the marble and the sheet and there was no air resistance it would just keep rolling forever along that sheet.

This is the way to think about this in terms of how objects with mass in the universe curve the space around them. The Sun in its position in space is like the baseball on this sheet, and the planets are like the marbles. The planets are just orbiting the Sun, in the curved space produced by the mass associated with the Sun. The reason the planets are not falling into the Sun is because unlike the marble, which had friction between it and the sheet and it would slow it down maybe and fall into the baseball in the case of the rubber sheet, in the case of the planets going around the sun, they're traveling in a near-vacuum of solar system space and there's nothing to slow them down; they just keep going. They're just orbiting the sun in the curved space produced by the mass associated with the Sun.

As we think about this, we note that objects in the Solar System—planets and asteroids—orbit the Sun following the curved space associated with the massive Sun; but not only do objects follow this curved space, but as light travels through this space, it will also bend to this curvature. Let's think about this: What this means, then, if light is coming toward us in the direction of the Sun from a star beyond the Sun, it will not move straight; it will be curved by the curvature of space associated with the mass of the Sun. it's this idea that helped Einstein come up with a key test of general relativity in 1916. What he said was: "If general relativity is right and space is indeed curved in the presence of mass, it should affect objects with mass and light such that if I looked at a star right on the edge of the Sun, that light should be curved. The Sun's gravity—the curvature of space associated with the Sun—would affect the apparent position of a star on the sky right at the sun's edge and shift its position by 1.7 arc seconds from its true sky position. Here was a fundamental, basic prediction of general relativity: If this theory is correct, astronomers should observe that when stars are very close to the sun, their positions on the sky will shift by 1.7 arc seconds. This was a wonderful test of general relativity; it's a hallmark of the best science. This is how science works: You come up with an idea—a hypothesis, a model, a theory—and based on this you make observable predictions, either through experiments on Earth or something you can determine by studying the heavens; and if the observations meet with the theory's expectation than the theory is supported, and indeed one can look for other tests of the theory. On the other hand, if the observations do not support the theory, the theory is wrong. It's important to realize just because Einstein says something is true, that doesn't mean it's true. Science doesn't play favorites. You come up with an idea, you come up with a prediction, and you test it. If it's verified, then the theory holds and you look for other ways to test it; but if it does not verify, then the theory is wrong. Here was a clear, fundamental test of general relatively; and if this shift was not observed, general relativity was wrong.

This test is hard, right? How could you measure the shift of a star right next to the Sun? How do you do that? The Sun's up during the day, and if the sky is bright it's hard to see a star anywhere in the sky but much less a star really close to the Sun; how could you even do this kind of an experiment to test general relativity? The good news is occasionally Earth experiences something called a "solar eclipse" when the Moon covers the solar disk and one can actually see stars close to the Sun. How does this work? A solar eclipse is one of the most fundamental, wonderful phenomena in nature; and at first glance, it seems kind of crazy because we know the Sun is huge— physically it's a big star—and the Moon is a moon orbiting the Earth; they're very different in physical size. But just by essentially a coincidence, the Sun is so much further away than the Moon that they both have the same angular size in the sky, half a degree; so it's possible for the Moon to cover the Sun.

The Moon orbits the Earth once every month or so; so you might say, "Oh, if the Moon's orbiting the Earth every month, then we should have a solar eclipse on Earth somewhere every month." Actually, we don't, because the orbit of the Moon around the Earth is inclined by an angle of five degrees with respect of the orbit of Earth around the Sun; so we don't get an eclipse of the Sun every month. It works out if you go through the celestial mechanics: We get a solar eclipse roughly about twice a year, and one can prodiot thooo oolar oclipcoc well in advance; you can predict when they will occur and where on Earth they will occur. The "where on Earth" is also a key thing to recognize about a solar eclipse: When a solar eclipse happens, it is not viewable by everybody on Earth; it only carves out a path of totality, we call it—a cut across the Earth—as this happens. What you have is a path no more than a few hundred kilometers wide in a wide swath thousands of kilometers across the Earth where you'll actually be in the right position to see the Moon pass directly in front of the Sun. At every point along this path of totality, the actual time at which the Moon covers the entire Sun is no more than seven minutes. If you're going to observe a solar eclipse, you have to be in a place that hopefully the weather will be good and you only get a short time to see it. But if you're ever seen a total solar eclipse, it's a fantastic phenomena; to actually have almost night occur on the Earth in the middle of the day is really an amazing phenomena to see, but it's a rare thing to see because it only happens about twice a year and only in very specific places on Earth that you can predict.

After Einstein made his prediction that here was a test of general relativity, here was a test of this crazy idea that space is curved—if you can observe a star near the edge of the Sun during a solar eclipse you can do this—after Einstein made this prediction in 1916, it was realized by a team of British astronomers that there would be an advantageous solar eclipse to test this in 1919. The path of totality would sweep across South America and the Atlantic Ocean and into Africa, and the British astronomers sent teams to both Africa and South America with the hope being if we have teams in two different places, we're less likely to have a problem with the weather or something else that would prevent us from making these measurements. The eclipse happened, the British teams were there, they observed it, and the amazing is they confirmed Einstein's prediction: They looked at a star right on the edge of the Sun and it indeed was shifted 1.7 arc seconds.

Let me emphasize this again: The way it works is you look at a position of stars on the sky when the Sun's in some other part of the sky and measure their relative positions, and then you measure their relative positions when the Sun is right there on the edge of one of those stars; and they noticed there was a shift. This was astonishing, because this changed the whole perception of reality long established by Newton hundreds of years earlier. The idea that space was curved, that general activity, this is the way the universe really worked? With this achievement, Einstein became a worldwide celebrity; he was famous already in physics circles for many other discoveries, but this achievement, to actually change the face of reality itself, that space does curve in the vicinity of massive objects, was just an amazing thing to behold. Since this test that was entirely successful, general relativity has passed all subsequent tests of its validity; indeed, general relativity is the way all physicists and astrophysicists deal with objects with a tremendous amount of mass in the universe. General relativity is the way our best idea to understand how the universe works and indeed we definitely need it when we talk about objects with a lot of mass in a small area.

The extent to which the space curvature around a massive object will make it a gravitational lens depends upon a number of factors including its mass, its size, its distance, its alignment with a background light source, and the distance of that source. The best way to see the effects of lensing can be by simulating the movement of a nearby, compact massive object across familiar backgrounds; let's start with moving such a compact massive object across the Baltimore city skyline. Imagine you're at some distance looking at the Baltimore city skyline and some compact massive object passes across your line of sight. In order to see this kind of effect, it has to be an amazingly compact, amazingly massive object. The reason the Sun wouldn't be a good choice is because the Sun's curvature—the space curvature associated with the Sun—is only about 1.7 arc seconds, a tiny little angle at its edge; and the reason it's so small is because although the Sun is very massive, it's huge, but if you take the mass of the Sun and scrunch it down into something a few hundred kilometers, a few kilometers in size, now you have more of a distortion of space, time, and its vicinity because not only is it massive, but it's very tiny. We look at this distortion of the background image of the Baltimore city skyline here; understand clearly, if there really was that compact massive object moving between your sightline and Baltimore, the entire Earth would be being ripped to shreds.

This isn't entirely realistic here in terms of what would really happen, but in terms of the view it is realistic, and what you're seeing here as the light waves come to you from Baltimore this compact massive object and the distorted space around it warps, distorts the image in the background of the city. In some places, as this object moves across, you see an increase as brightness; and the light rays that normally wouldn't hit your eye are brought into your eye. In other places, it leads to just a total distortion of the image. In many ways, this image looks like something you might see if you went into a carnival in some funhouse with warped mirrors, just seeing these kinds of distortions; that's what it can affect, if you have a massive object particularly compact, it can distort the space in its vicinity.

You can see a similar effect if you have a galaxy field; here we have a simulated galaxy field and we're taking that same compact, massive object and it's passing near to us so it's distorting the background. Imagine you went out at night and you're looking at the sky, and you saw weird distortions in the sky: stars changing brightness, fuzzy things changing in size; all kinds of things happening slowly as this object moves across. You would know that something very compact and very massive was moving very close to you. If you ever see this, you're almost certainly due for a very bad evening, because this would be something that would only happen if you had really a massive compact object pass close to Earth. The good news once again is on a list of cosmic things to worry about, something that compact and massive passing close to Earth is extremely unlikely.

The simplest case of a gravitational lens is when we have the observer, the lens, and the background object all perfectly aligned. By "perfectly aligned," I mean within much less than an arc second. The result of that is an Einstein Ring. What you see is the image of the background object is lensed by the lens; the space curvature around this lens takes the light and spreads it into a ring of light right around the object that's doing the lensing. Most likely you won't get a perfect alignment; and then you won't see a complete ring, you'll just see multiple, arc-like images of a lensed object. Do we see these things? Do we actually see these Einstein Rings? In the case of star-on-star alignments in the Milky Way Galaxy, they don't yield Einstein Rings, and the reason is the stars typically lack the mass and the angular size to produce resolvable Einstein Rings; I mean resolvable even with the Hubble Space Telescope. It's there, but it's such a tiny angle, no optical telescope, even Hubble, can pull it out. But the brightening associated with that little Einstein Ring, we can detect the star actually brightening.

This is a phenomenon called "gravitational microlensing." If you're observing a rich starfield that extends way off into the distance—maybe something like the Sagittarius Star Cloud—and you're looking at one of those nearby stars, you actually have a star-on-star alignment, you may watch one of these stars actually slowly increase in brightness and then over time slowly fade away on time scales of a few hours, days, months. You're actually seeing this gravitational microlensing, and this is certainly interesting from the context of detecting planets around other stars; believe it or not, a few planets around other stars have been seen though this kind of gravitational microlensing effect, where a star gets amplified due to the gravitational lensing effect as a foreground star amplifies the light from a background star. This phenomenon has been seen in terms of the star-on-stars in the galaxy in certain directions where there are lots of stars over great distances, like the Sagittarius Star Cloud. In terms of actually resolving an Einstein Ring, actually seeing the ring of light around a particular lens, all those that have been seen to date have involved galaxy-on-galaxy alignments, and about 50 partial to full Einstein Rings have actually been discovered optically today. In these cases, the typical Einstein Ring diameter is on the order of a few arc seconds.

The images you can obtain with Hubble—the high-resolution images—are absolutely key to understanding the physics associated with these Einstein Rings. What you can do with Hubble, because you can see in such great detail, you can actually establish as you study an Einstein Ring the morphology of the galaxy being imaged. You can also estimate better: What's the mass of the galaxy doing the lensing? Hubble also has the remarkable achievement that it's actually imaged the Double Einstein Ring, which is an alignment of not just two but three galaxies along the sightline. You have a massive lens galaxy at 3 billion light-years, and then you have lens galaxies at 6 and 11 billion light-years. Just lining these all up is like not just getting a hole in one during a round of golf but getting a hole in one on back to back holes in a round of golf. This lensing brightens these very distant galaxies; it not only creates the ring, but it amplifies them, makes them much brighter than they would otherwise be. Indeed, these other galaxies, 6 and 11 billion light-years away, would be very hard to detect even with Hubble. Hubble, through these kinds of observations, has revealed a number of individual lenses, and in the portfolio here you see in this slide you can see that the lensed images include a variety of patterns: some are rings, some are pieces of rings, some are multiple images. Hubble's ability to image Einstein Rings and study them well comes specifically from its ability to study objects at such high angular resolution on the sky.

The most spectacular cases of gravitational lensing involve distant rich clusters of galaxies. These clusters have a vast amount of mass and they're bigger on the sky than an individual galaxy; thus they're able to curve the space in their vicinity over a wider sky area and in a more complex manner than does a single galaxy. Thus, these clusters can produce many lensed images depending on the richness and the distance of background galaxies. Such is the case for the rich galaxy cluster Abell 2218 at a distance of 2 billion light-years. The Hubble image of Abell 2218 reveals about 80 bright cluster galaxies and over 100 gravitationally-lensed arc-like images of the background galaxies. This image is a composite of blue, yellow, and red images, and in this composite the color you see roughly approximates what you would see with your eye. The total amount of time put into this image is a little over nine hours, and it spans about 1/11th the width of the full moon. At the distance of Abell 2218, this sky width corresponds to a physical distance of about one-and-a-half million light-years.

The thing when you stare at this imagine, this is relativity right here; you can actually see in this image, you can see the curvature of space. It's amazing what Hubble has been able to do in high resolution and see these thin arcs; this is just like a living embodiment of what general relativity is all about, you can see the curvature of space in the form of all these curved images of distant galaxies. Indeed, many of these arcs seem to encircle the core elliptical. However, if you look carefully at this image, the overall arc pattern is not symmetric; it indicates complex, large-scale space curvature. When you study some of the arcs in detail, you note that some of the faintest arcs are very red indicating that they are likely to be a great distance at high redshift. Indeed, in 2002, it was possible to obtain brown-based spectra of two of the faint red spots in this image, and they yielded a very high redshift. The galaxy responsible for these two red spots was 13 billion light-years away, and it turned out the galaxy responsible for them was indeed a young galaxy, actually much smaller than the Milky Way. The lensing provided by Abell 2218 brightened it by a factor of 30. This image has also revealed other such distant galaxies. The important point to recognize here: Without the brightness boost by the gravitational lensing and the sharp eye of HST to see these thin arcs, these galaxies would have been almost impossible to detect.

By studying the positions and shapes of the over 100 arcs in this image, it's possible to construct a map of the lensing mass responsible for the space curvature producing the arc distortions. In other words, you're making a gravitational map of all the mass in this cluster of galaxies. You can sort this out by studying all these lenses. But this is just one mass estimate; there are other ways to get estimates of the mass in the cluster. For example, if you measure the brightnesses and the colors of the light from the galaxies, you can estimate the amount of mass in this cluster that's in stars. The light itself, based on what we know about stars in our galaxy, and its color can tell us roughly how much mass is tied up in stars. Furthermore, if we study the gas between the galaxies in this cluster, we see that it glows in x-rays; in other words, there's lots of hot, tenuous gas between the galaxies and Abell 2218, and indeed between the galaxies and many rich clusters of galaxies.

Where does this hot gas come from? At earlier times, when these galaxies were interacting with each other and colliding, some of the gas that did not go into starburst was stripped and expelled into the space between the galaxies and clusters and that gas was very hot; and since it's so tenuous, it takes many, many years to radiate it away slowly, these energies and x-rays. From these x-rays, we can estimate how much mass is in the gas between the galaxies. The point is: We have the mass between the galaxies from the x-rays, we have the optical light and other measures to get the light of stars and galaxies, so when we add up all the mass it should equal the amount indicated by the lensing in terms of the gravity, right? It should be the same, more or less. It's not; the lensing mass is about 10 times greater than the mass in the stars and the gas; and that holds true not just for Abell 2218 but other clusters that have been studied through gravitational lensing. What this is telling us is that most of the mass in Abell 2218 must be dark matter. The nature of this dark matter is unknown. This dark matter is basically defined as matter that interacts with visible matter through gravity but not through electromagnetic forces. Without these kinds of electromagnetic interactions, dark matter cannot reveal itself by emitting or absorbing photons. Thus, we can't "see" what the dark matter is; but we can find that it's there through effects such as gravitational lensing.

It turns out that the gravitational evidence of dark matter is not just confined to Abell 2218. It's found in galaxy clusters and indeed individual galaxies throughout the universe. In particular, Hubble observations of gravitational lenses in other galaxy clusters are continuing to provide new clues about the character of this dark matter. For example, the Hubble image of the massive galaxy cluster Abell 1689 at a distance a little over 2 billion light-years away shows us hundreds of lensed background galaxies. If we do close-ups on various sections of this culture, we see a veritable zoo of lensed images. We see arcs that are thin, some that are thick; we see some arcs that are short, some that are long; we see arcs of different colors. What you can do with such a rich variety of arcs over a relatively wide chunk of the sky— remember with these clusters you can see the effective curved space over arc minutes of sky as opposed to like a single Einstein Ring, which is only a few arc seconds—in such a way by studying these gravitational lenses across a galaxy cluster, you can model the observed lensing pattern over a significant chunk of the sky. And you can answer questions like: What's the distribution of the dark matter over these few arc minutes across the cluster? Is it clumped around individual cluster galaxies, or is it distributed more smoothly across the cluster? In the case of Abell 1689, the pattern of these lenses shows us that the dark matter is distributed somewhat more smoothly than the galaxies' light but with evidence of larger-scale structure. In effect, the lensing is allowing us to take a "gravitational photo" of the dark matter in the cluster since remember, most of the matter we're seeing, most of the gravity, is associated with the dark matter as opposed to the visible matter and the gas.

There's strong evidence that at least some of the dark matter is associated with individual galaxies. It's been known since the 1970s, for example, that the dark matter is likely to be common in spiral galaxies, including the Milky Way, based on the rotational velocities of its outer stars. The original expectation of how mass would be distributed in spiral galaxies like the Milky Way—or M33, as shown in this slide—was that light traces mass, and so as we look toward the edge of a spiral galaxy there should be less mass because we are seeing fewer stars. Indeed, in that pattern what you'd expect is that the velocities of stars as we went further and further and further away from the center of the galaxy should decrease. The analogy would be how planets orbit the Sun. In the case of the solar system, most of the mass— almost all of it—is concentrated in the Sun—that's where the mass really arises in the solar system—and we notice, of course, that the outer planets orbit slower than the inner ones.

The observed velocities would be expected, when we look at a galaxy like M33, to go down just as we see in the solar system, but what's actually observed is we see a constant velocity out to distances where we don't even see stars in the galaxy, or a slow rise. These are called "flat rotation curves"; that's a characteristic not just of the Milky Way, but almost all the spirals we see in the sky; these flat rotation curves. How do we understand them? It turns out they are best understood in terms of a dark matter halo; that spirals like the Milky Way, M33, and others have this huge dark matter halo that extends to much greater distances than the size of the optical disk itself, and this dark matter halo has much more mass than the visible matter in the disk. What is this dark matter made out of? Given its vast abundance, it's almost certain that this dark matter doesn't consist of things like protons and neutrons that make up you, me, planets, stars, that kind of stuff; it has to be something else. There simply isn't enough matter like we're made out of in the universe to explain this huge amount of matter. The best guess is that this dark matter is actually some kind of massive particle formed early in the history of the universe that doesn't interact much at all with regular matter—some have dubbed this "weakly interacting massive particles"—and they're very, very hard to detect for the simple reason they don't interact with regular matter. But a lot of astrophysicists and a lot of physicists are looking hard for this dark matter, and there is some hope that sometime soon we'll actually understand what this is; indeed, it's far more likely we'll understand what this dark matter is before we ever figure out what dark energy is.

In summary, it's ironic that through gravitational lensing, the dark matter in the universe can provide illuminating views of very distant galaxies. One of the first Hubble images taken after the 2009 servicing mission was that of the galaxy cluster Abell 370. Among the hundreds of lensed galaxies in this image, one stands out. It's a well-resolved, distant spiral with a long tail. At first glance you might think it's due to a collision like the Tadpole Galaxy; but no, it's a lensing artifact produced primarily by the dark matter in this foreground galaxy cluster. This lensed image makes it possible for Hubble to study in detail a galaxy at a distance of 9 billion light-years. Hubble images such as this one and that of Abell 2218 illustrate that we don't have to know the composition of dark matter in order to interpret and take advantage of its key contribution to gravitational lensing in the vicinity of distant galaxy clusters. Next time, we're going to go as deep as we can go with Hubble and probe the Hubble Ultra Deep Field image and study how far HST can see without the aid of gravitational lensing. Please join us then.

The Hubble Ultra Deep Field
Lecture 11

Indeed, there are over three times as many galaxies in the Ultra Deep Field than the number of naked-eye stars in the entire night sky. Furthermore, all of these galaxies fill a sky area that's 60 times less than that of the full Moon.

Following Edwin Hubble's discovery of the expanding universe, astronomers developed theories to explain the phenomenon. The Big Bang idea propounds a singular origin in the past for everything and expects that, over time, galaxy density will decrease. As the galaxies pull away from one another, the voids will just keep expanding. Until 1965, no evidence definitively favored this idea, but then Arno Penzias and Robert Wilson serendipitously discovered the **cosmic microwave background radiation** and essentially settled the debate in favor of the Big Bang. At Bell Labs in Holmdel, NJ, Penzias and Wilson were testing microwave receivers for satellite communication and detected microwave background "noise" that seemed to come from everywhere. At about the same time, it just so happened that astrophysicists at Princeton University were thinking that it might be possible to detect remnant radiation from the time shortly after the Big Bang in the form of a microwave background across the sky. These two scientific groups published joint papers in the *Astrophysical Journal* in 1965, and Penzias and Wilson were later awarded the Nobel Prize for discovering the signal of the Big Bang.

In the standard Big Bang model, the microwave background dates back 13.7 billion years; it is the redshifted remnant from a time when the entire universe was small, dense, and hot like a star. The universe was a vast plasma of photons bouncing off electrons. Since then, with the expansion of the universe, these photons have redshifted by a factor of 1,000 from optical to microwave wavelengths. Since Penzias's and Wilson's discovery of the microwave background radiation, a long series of microwave background measurements of increasing sensitivity have been made in an effort to pick up tiny temperature fluctuations in the radiation across the sky. These temperature fluctuations reflect small matter fluctuations in the universe that could indicate galaxy formations long ago. The most sensitive all-sky microwave background measurements to date, those taken by the Wilkinson Microwave Anisotropy Probe, have revealed temperature variations of 1 part in 100,000 across the sky. Computational astrophysics uses these data to create models of the early universe that are tweaked as more data become available.

Can we look back far enough with our telescopes to see galaxies actually evolving over time, as we would expect in a Big Bang model? In other words, if the Big Bang model is indeed correct, the universe should evolve over time, and the most distant galaxies should be morphologically distinct from the present-day galaxy population. Can the Hubble's exceptional sky resolution resolve the most distant galaxies to determine whether their morphologies differ from those of the spiral and elliptical galaxies that we see today? As it turns out, the 1996 Hubble Deep Field image showed that the most distant galaxies tend to be peculiar, which is consistent with the evolving universe expectations of the Big Bang model.

> **[T]he 1996 Hubble Deep Field image showed that the most distant galaxies tend to be peculiar, which is consistent with the evolving universe expectations of the Big Bang model.**

In 2004, an even more ambitious effort, the Ultra Deep Field, captured objects 4 billion times fainter than the faintest object you can see in the night sky with your eye. When we study this image in detail, we see 10,000 galaxies spread out over a sky area equivalent to only about 1.7 percent of the full Moon sky area. If you assume this is an average galaxy density, then the entire sky would yield a population of 130 billion galaxies in the observable universe. Like the Hubble Deep Field image, close-ups of the Ultra Deep Field reveal many galaxies with asymmetric morphologies; in particular, 165 tadpole-shaped galaxies tend to be appreciably smaller than the Milky Way, suggesting that galaxies grew bigger through frequent collisions. The reddest Ultra Deep Field galaxies are the most distant dwarf galaxies, and they date back to a time less than 1 billion years after the Big Bang. Their star formation rates are about 10 times greater than those of nearby galaxies, which makes sense. If there were numerous collisions going on in the early universe, we'd expect large starbursts to occur, even in interactions among small galaxies.

What about the very first stars and the very first galaxies? If we can push Hubble any further, can we see those? Unfortunately, the very first stars and galaxies in the universe are probably too faint for Hubble. Their light is probably redshifted into the infrared part of the spectrum, where Hubble is not sensitive. One of the things that the Hubble Ultra Deep Field shows us is that even Hubble has its limits; we can't see all the first stars and first galaxies in the universe. ■

Important Term

cosmic microwave background (CMB) radiation: The redshifted radiation from a very early time (380,000 years after the Big Bang) when the expanding universe transitioned from its hot, dense, bright origin to a cooler, transparent state.

Suggested Reading

Beckwith et al., "The Hubble Ultra Deep Field."

Gardner, "Finding the First Galaxies."

Singh, *Big Bang*.

Questions to Consider

1. What would the Hubble Ultra Deep Field look like in a Steady State universe?

2. What are all the factors involved in choosing a sky region for the deepest possible image of the most distant galaxies?

The Hubble Ultra Deep Field
Lecture 11—Transcript

Welcome back to our tour through the universe of the Hubble Space Telescope. Last time, we saw how Hubble has made observations of curved space routine and has utilized gravitational lensing to study dark matter and a sample of very distant galaxies. The unparalleled ability of Hubble to resolve and image distant galaxies without the aid of gravitational lensing is also important since only a very tiny fraction of the distant sky is amplified by lensing. The corresponding question is: How far can we see with Hubble?

This is a question that could be asked of any telescope, or even the human eye. In the case of the naked eye, the most distant galaxy that can be seen by most of us under optimal dark sky conditions is the Local Group spiral M33, which at a distance of 3 million light-years. As we employ ever larger ground-based telescopes and digital cameras to improve our vision, it's possible to actually detect the optical light of galaxies from the ground over 10 billion light-years away. At these distances, we're looking not only far, far away, but also deep into the distant past. Indeed, by observing the universe of galaxies as it was so long ago, we can potentially address key issues concerning the evolution of galaxies in the universe. Fortunately, Hubble has the necessary ability to not only detect galaxies beyond 10 billion light-years, but also to image their optical structure with a resolution that is unattainable from the ground.

This ability has been realized in the form of the Hubble Ultra Deep Field, the deepest optical image of the cosmos made to date. In terms of both science and beauty, there's absolutely no question that the Hubble Ultra Deep Field is a top 10 Hubble image. Almost every point of light in this image is a galaxy far beyond the Milky Way. Indeed, there are over three times as many galaxies in the Ultra Deep Field than the number of naked eye stars in the entire night sky. Furthermore, all of these galaxies fill a sky area that's 60 times less than that of the full Moon. Although it's small in sky area, the Ultra Deep Field is a vast core sample in time and space ranging from its nearest star 2,000 light-years away to its most distant galaxy at 13 billion light-years. In today's lecture, we're going to discuss the Ultra Deep Field in the context of galaxy evolution and the nature of our expanding universe.

This discussion involves cosmology. Cosmology is the area of astronomy dedicated to the study of the origin, evolution, and fate of the universe as a whole; essentially all that is, all that was, and all that ever will be. Prior to Edwin Hubble's discovery of the expanding universe, cosmology was essentially a data-starved field. But by the 1950s, the observations of the expanding universe had led to two competing cosmological theories to explain the universe: the Steady State and the Big Bang models. Let's contrast these two models: The Steady State universe is essentially infinite in time and space. A Steady State universe had no beginning, it will have no end, and it's everywhere; infinite. In a Steady State universe, it will maintain a constant galaxy density over time; it has to do it if it's Steady State. But we already know that the universe is expanding, so how does a Steady State accomplish this? As the clusters of galaxies pull away from one another in a Steady State universe, new galaxies are created in the voids that open up as these clusters of galaxies separate. A Big Bang universe is quite different; a Big Bang universe had a singular origin in the past, and a way to characterize a Big Bang universe is that over time, its galaxy density will decrease. As the galaxies pull away from one another, the voids will just keep expanding; and that will result in a net decrease in density in the Big Bang universe over time. Note a key contrast in terms of the creation of matter: In a Steady State universe, as we said, there will be a continuous creation of matter throughout the history of the Steady State universe. In the case of the Big Bang, matter is created only once in a Big Bang universe, at the moment that in which a Big Bang universe originates. Note there's a key observational difference between these two models: A Big Bang universe of galaxies should evolve over time as the density of galaxies in such a universe decreases; there should be no such evolution in a Steady State universe.

The debate between the Steady State and the Big Bang models to explain our universe continued into the 1960s simply because there was no convincing piece of evidence to definitively favor one idea over the other. Certainly the expanding universe observations seemed to be more simply explained in terms of the Big Bang, but that wasn't a definitive piece of evidence. The situation changed, however, in 1965 when Arno Penzias and Robert Wilson serendipitously discovered the cosmic microwave background radiation and essentially settled the cosmology debate in favor of the Big Bang. Let's spend a few minutes talking about this, because it's really important to understand this microwave background radiation in the context of the evolution of galaxies in the universe. Penzias and Wilson were research scientists working at Bell Labs in Holmdel, NJ, and they were testing microwave receivers for satellite communication using a 20-foot horned microwave antenna. This was important at the time; this was the 1960s, this was the heyday of putting satellites and people in space. It was very important for communicating with people in orbit to have very efficient, sensitive communications at microwave wavelengths to communicate with satellites and astronauts. We needed to develop the most sensitive receivers we could, and that's what Penzias and Wilson were working on.

In studying these receivers and pushing them to the limits, they detected this microwave background "noise" all over the sky; whenever they pointed their horned telescope, there would always be this little background microwave signal. They were scratching their heads; they couldn't figure out: Where was this background hiss coming from? First they thought, as you usually do when you detect something you don't understand, you see what's in your instrument. So they checked their instruments, they checked these receivers, and they actually crawled inside this 20 foot horned antenna and they found a lot of pigeon droppings and thought, "Maybe could this somehow be due to pigeon droppings?" They scraped out all the pigeon droppings but no, the noise was still there. So they said it doesn't appear to be in the telescope, it doesn't appear to be in the receivers, it must be somewhere in the sky. But it was amazing, because this radiation, this hiss, was the same everywhere in the sky; furthermore, it showed no daily, or weekly, or monthly variations in intensity. Given that, it was pretty obvious it almost certainly wasn't due to out a source in the solar system, because it probably would have had some variation over time. Also, another reason to rule out being due to something in the solar system was that if it was due to something associated with the planets or the asteroids, the planets and the asteroids in the asteroid belt are concentrated in the ecliptic plane orbiting the Sun; so the ecliptic plane on the sky, if that was due to something in the solar system, one would expect to see an anisotropy along the ecliptic plane. What do I mean by an "anisotropy?" It's that along the ecliptic plane it might be brighter in this microwave hiss, it would be a little bit more intense; but it wasn't. Furthermore, once they ruled out a solar system source, Penzias and Wilson could rule out a galactic source. If it was due to something in the galaxy, it would almost certainly be concentrated along the galactic plane where most of the stars are and the dust and the gas; in other words, along the Milky Way across the sky. But no, this hiss was not any brighter along the Milky Way; no anisotropy along the Galactic plane. Indeed, it was very isotropic across the whole sky; essentially the same intensity of this radiation wherever they looked. They were stumped. They could only guess it was something on a cosmic scale, beyond the Milky Way galaxy.

It just so happened that a few miles down the road a group of astrophysicists at Princeton University were thinking about the Big Bang and they were trying to understand the conditions that would have existed in the universe shortly after the Big Bang. They realized that under these conditions we might today see a remnant radiation from that time in the form of a microwave background across the sky. They actually began to build equipment to look for this; but before they finished building their equipment—receivers and the telescope to see this—they made contact, they found out about Penzias's and Wilson's discovery, and voila, they found it; these two groups hooked up and it was realized that Penzias and Wilson had actually discovered the signal of the Big Bang. They published joint papers in the *Astrophysical Journal* in 1965, and Penzias and Wilson were later awarded the Nobel Prize for this amazing discovery.

With this microwave background interpreted as cosmic in origin and attached to the Big Bang, did the Steady State universe have a counter argument? No, the Steady State universe had no simple explanation at all for this microwave background radiation. In the standard Big Bang model, though, this microwave background dates back 13.7 billion years; it dates back to a time when the entire universe was small, dense, and hot like a star. The microwave background radiation is the redshifted remnant from this very early time. Let's spend some time and take a detailed look at what the universe was like during these early times. In the standard Big Bang model, during the first 380,000 years, the entire universe was as bright as the inside of the Sun; the entire universe was light; wherever you went it was bright. The entire universe was light. What the universe was made of at this time was a very hot, dense gas of mostly ionized hydrogen. When I say "ionized hydrogen," of course, that means protons, and in addition the electrons were there; so the universe consisted of protons and electrons, and lots of energetic photons carrying the energy in this very hot early universe.

Under these conditions of such a hot plasma, the photons in the universe would scatter off the electrons; that's why the universe appeared bright, because the photons were constantly bouncing off the electrons all over the place, and so wherever you looked in the universe during these times, photons would go in your eyeball and you'd see light everywhere. But it's important to remember what the universe is doing at this point in time: It's expanding; and as the universe continues to expand, it will cool. You might ask: Why does the hydrogen stay ionized during these early times? Because the photons have enough energy so whenever an electron and a proton combine to form a hydrogen atom, an energetic photon would come along and break them back up into a proton and an electron. But over time, as the universe continues to expand, the photons in the universe—their wavelengths—get stretched with the expansion of the universe and they lose energy and the universe cools off; and eventually, by the time the universe reaches its 380,000th birthday, these photons have cooled down to a temperature of about 3,000K, and at this temperature the photons no longer have energy to break apart an electron and a proton if they merge together into a hydrogen atom. So at this point in time, 380,000 years after the Big Bang, the electrons and the protons start recombining, the photons can't break them up, and the universe goes neutral; we have hydrogen atoms. Now the photons, which were bouncing off the free electrons before, don't have any electrons to go bounce off of anymore, and they then go streaming through the gas. Essentially what happens then is the universe becomes transparent as the photons stream through the now-neutral gas of the universe.

Since this time, what we call the Recombination Era, the universe has expanded by a factor of 1,000. These photons, when the temperature was about 3,000K and these photons typically had wavelengths around optical wavelengths, they've been redshifted by a factor of 1,000; their wavelengths have been stretched by a factor of 1,000 from optical to microwave wavelengths. That's how we see them today as microwave photons; these very same photons that were optical photons long, long ago when the Recombination Era occurred in the universe. Thus, the observed microwave background actually provides us the earliest "baby picture" of the universe, when it was only 380,000 years old. This is a very important point to understand: We can't see back any earlier than this time, and we can't see the Big Bang itself, because the universe was opaquely bright during times before 380,000 years before the Big Bang. An analogy would be that we can observe the bottom surface of a cloud deck on a cloudy day, but we can't see into the cloud or beyond the cloud to the Sun itself.

Let me stress this point with another way to think about it. Imagine you're out at night looking at the sky, and you can only see so far admittedly with your eye, but travel with your mind's eye and think about what you're seeing. As you look out into the darkness of deep space and you go past the solar system—which is typically light-minutes to light-hours in dimension—you go beyond it and you get light-years away, and tens of light-years, and hundreds of light-years; you're getting into the Milky Way galaxy (let's assume you're looking through the halo of our Milky Way), and then you get tens of thousands of light-years, and hundreds of thousands of light-years away. Now you're getting into our Local Group of galaxies, and then you start getting millions of light-years, and tens of millions of light-years, and a hundred million light-years away. Now you're looking at our local universe, more distant galaxies, past the Virgo Cluster. Your sight continues; with your mind's eye you keep going deeper and deeper, and you're going further and further into the past along with going far, far away. Now you get to billions of light-years away, and now you're getting to such great distances, into such great look back times in the past, the photons from these early eras are getting redshifted from optical into infrared. Then continue with your mind's eye: If you go beyond the most distant galaxies, the very earliest galaxies and beyond, eventually you see the microwave background radiation; that's a wall on the sky. You're looking back to a time when it was 380,000 years after the Big Bang, when the universe everywhere was as bright as the Sun. But you can't see beyond that; even though theoretically in some sense we could, if you looked deep enough with a big enough telescope, see the moment of creation itself, see the Big Bang itself, you can't see it because it's shrouded by these early times when the universe was as bright as the Sun.

Since Penzias's and Wilson's discovery of the microwave background radiation, there's been a long series of microwave background measurements of increasing sensitivity in an effort to determine the degree of isotropy in this surface of last scattering. Why is this important; why do astronomers want to do this? The point is, what I mean by "degree of isotropy" is, as we already talked about since it's cosmic in origin, at some level we expect this microwave background radiation to be uniform across the whole sky. But the very seeds of the galaxies, the large-scale structure of galaxies and clusters of galaxies we see today, should be in this earliest baby picture of the universe; there should be tiny little temperature fluctuations on tiny scales in this radiation across the sky that's reflecting little matter fluctuations in the universe, and those little matter fluctuations over time will produce the universe of galaxies we see today. If these tiny little matter fluctuations could be observed—these little anisotropies in this microwave background radiation—we could perhaps understand how the galaxies we see today have evolved.

This is why it's so important to try to detect these; and indeed, in the most sensitive all-sky microwave background measurements to date, the Wilkinson Microwave Anisotropy Probe (or WMAP)—this is a satellite—has revealed tiny temperature variations, actually seen these effects, seen these tiny little temperature fluctuations, 1 part in 100,000 across the sky. Here we see this all-sky map of what it looks like in microwave at very high precision. For the most part, remember the radiation is amazingly isotropic; but when you look at this sensitivity you can see little warm regions that are red on this map or yellowish and little cool regions that are bluish in this map. How do we deal with these fluctuations? How do we make sense out of them? What we can do is using this as initial conditions, with this image we know what the universe looked like 380,000 years after the Big Bang. We also know from our study of galaxies—nearby galaxies, distant galaxies—what the universe looks like now and indeed back to billions years ago in terms of the distribution of galaxies on the sky.

What we can do, then, is take this earliest baby picture and then let our computational astrophysics friend go to work; develop models of the universe that take into account the expansion of the universe, the gravity between the galaxies and these fluctuations, dark matter, dark energy. You put everything and the kitchen sink in these models, and then you put them into computers and you crank out SIM universes. In other words, given the initial conditions of the anisotropies in the Big Bang, can we reproduce the galaxies we see today by putting in all of physics and letting our best, most powerful supercomputers go to work? In so doing, what we find is that these models more or less reproduce the large-scale structure of the galaxies we see today. Once again, there's a synergy between the models and the observations: As we get better observations, the models improve; as the models improve and ask more questions, we're driven to get even better observations. One of the key things to do for observational astronomers is to find better data on the most distant galaxies we can find. The Holy Grail, indeed, would be to detect the very first stars and the very first galaxies in the universe. But also a very important step is: Can we look back far enough with our telescopes to see galaxies actually evolving after time, as we would expect in a Big Bang model? In other words, if the Big Bang model is indeed correct, the universe should evolve over time and the most distant galaxies should be morphologically distinct from the present-day galaxy population. No telescope is better equipped currently to test this hypothesis than Hubble. Indeed, we need Hubble; we need Hubble's very exceptional sky resolution to not only detect the most distant galaxies but actually resolve them well enough to see is their morphology different than what we see today? Do they actually look different than spiral galaxies and elliptical galaxies?

In 1996, Hubble completed the Hubble Deep Field, its initial attempt to push the limits in imaging distant galaxies. There's an interesting story behind this image; I want to spend a few minutes talking about it because it's related to how astronomers obtain time with the Space Telescope and other aspects of the Hubble project in terms of exactly how astronomers deal with their data and you end up with exciting projects like the Hubble Deep Field Project. Actually, the Deep Field Project was instigated by then Bob Williams, who was then the Director of the Space Telescope Science Institute. The Space Telescope Science Institute director administers the operations, the science operations, of Hubble. What Bob Williams did is he decided to dedicate 150 orbits of his discretionary telescope time to this Hubble Deep Field Project. I should say an orbit of time; usually that's like the currency of Hubble observing among astronomers. An orbit, remember, is 97 minutes—it takes Hubble to go around the Earth 97 minutes—so an orbit typically involves the amount of actual on-star, on-galaxy observing time anywhere between 50 to 97 minutes, depending if the Earth is in your way during the orbit.

Among the telescope time that is awarded for Hubble that is available every year, up to 10 percent of it is up to the director's discretion; the director can decide how he or she wants to spend this time. What does this mean exactly? Let's look at the other 90 percent first. Most of this 90 percent of Hubble time is awarded competitively among astronomers, and every year astronomers submit about 1,000 proposals. Which proposals among these are actually awarded time, typically the criteria the selection committee looks at is certainly it needs to be extremely important science, number one; number two, it has to be science that can only be done with Hubble Space Telescope (why waste Hubble time if it can be done from the ground?). Furthermore, what the committees are looking for are proposals with clear scientific objectives. In other words, if you give us this time, the observations will mean either this or this in terms of addressing a particular scientific problem. What happens is that out of these some thousand proposals, only one out of seven is typically awarded time. That doesn't mean only one out of seven is worthy of time—easily more than half o the proposals are worthy of time; it's really hard serving on these committees and sorting out these excellent proposals—it's just that Hubble is such a precious commodity, we just can't afford to give out as much time as we'd like. Another key thing to recognize: Since the process is so thorough, there's a long time between the proposal and the observation, and it can be as long as two years. So the director's discretionary time is typically available, though, for very urgent projects— something that's come up on a new burst of some supernova or something else—or a risky project that might be looked at as this is just too risky in terms of actually working or the science is kind of just a little bit sketchy but perhaps important.

What Bob Williams decided to do is be really bold with his director's discretionary time. He said why don't we just sink these 150 orbits and take a very long exposure of one sky field? Just point Hubble and integrate away. The simple justification was: We're just going to look as deep as possible, and undoubtedly we're going to find something really important and really neat. You'd say: Wait a minute, that's not like a really focused proposal, you don't know exactly what you're going to find; is this really what we should be doing? But the point is what science has taught us again and again, often particularly if you have new technology, new equipment, pushing it to its limit you serendipitously often discover things that are more interesting than the science you planned; indeed, the cosmic microwave background radiation itself is a good example of that. What Williams decided to do in addition to pushing Hubble to its limits and taking really deep exposures was also make the image available to all astronomers immediately upon receipt. Usually Hubble observers get about a year proprietary rights for their own program. Despite that there's some criticism at the time about this approach—a "fishing expedition" type criticism—all that evaporated after Hubble Deep Field results came out.

Here we see the Hubble Deep Field image. It reveals 3,000 galaxies over a sky area equivalent to about .7 percent that of the full Moon. It is a 98 hour composite of 276 Widefield Planetary Camera Two images taken through blue, red, and near-infrared filters. This region for the Hubble Deep Field is located near the Big Dipper, and if you looked at it with your eye, or even a small telescope, you'd see nothing; it's blank. But with Hubble, and you let Hubble integrate for a long time, you see an amazing zoo of galaxies; and indeed, when the Hubble Deep Field image is studied up close, we see that it contains many peculiar galaxies. Indeed, if we follow up work on the Hubble Deep Field image, they show that these peculiar galaxies tend to be the most distant galaxies, which is consistent with the evolving universe expectations one has for the Big Bang. The fact that the Hubble Deep Field image has already led to more than 175 published research papers indicates that it's quite an amazing scientific bonanza from one man's fishing expedition.

Given the wonderful success of the Hubble Deep Field, an even more ambitious effort was launched in 2004, utilizing the more sensitive ACS camera (Advanced Camera for Surveys) onboard the Hubble. In terms of its faintest detections, the resulting Ultra Deep Field is about four times deeper than the Hubble Deep Field. Indeed, the faintest object in the Ultra Deep Field image is about four billion times fainter than the faintest object you can see in the night sky with your eye. Once again, if we looked at the Ultra Deep Field sky region with the naked eye, we'd see nothing, it's devoid of bright stars and galaxies, and it's located in the constellation Fornax. This image is somewhat larger on the sky than the Hubble Deep Field—it's about twice as large as the Hubble Deep Field—and here this composite image for the Hubble Ultra Deep Field involves images taken through blue, green, red, and near-infrared filters. It's a 268 hour net exposure involving 808

ACS images. It took 400 Hubble orbits to produce effectively this single image; 400 Hubble orbits are essentially 10 percent of the annual science total available at Hubble. When we study this image in detail, we see 10,000 galaxies spread out over a sky area equivalent to only about 1.7 percent of the full Moon sky area. If you took this—assume this is an average galaxy density—Ultra Deep Field galaxy density and spread it over the entire sky, it would yield a population of 130 billion galaxies in the observable universe.

Just as in the Hubble Deep Field image, close-ups of the Ultra Deep Field reveal many galaxies with morphologies unlike spirals and ellipticals. Even without the detailed spectroscopy of these Ultra Deep Field galaxies, it's possible to measure their colors from the images taken through the different filters and using those, we can separate out the most redshifted distant galaxies from those relatively nearby. This is pretty straightforward; if you see a galaxy that doesn't appear in the blue filter, doesn't appear in images with green, but it's only in the red or the near-infrared, you say that galaxy has to be pretty far away if its light is redshifted so much we don't see it at blue or green wavelengths; so given that, you can separate out to some extent the most distant galaxies here. What you find in this sort of analysis is that the most distant galaxies typically appear smaller and more asymmetric as compared to the spirals and ellipticals in our local universe. Among these, 165 tadpole-shaped galaxies have been identified in the Ultra Deep Field. These tend to be appreciably smaller than the Milky Way; and what this tells us looking at these galaxies is that among the first galaxies to form in the universe were smaller galaxies, and through epochs of collisions—remember the universe was a smaller, denser place then—through frequent collisions, they eventually grew into bigger galaxies, which is indicative indeed that there was an early epoch of mergers and acquisitions in the universe. The most red Ultra Deep Field galaxies are the most distant little dwarf galaxies, and they date back to a time less than a billion years after the Big Bang. Their star formation rates are about 10 times greater than those of nearby galaxies; which makes sense, right? If we have a lot of collisions going on in the early universe, we'd expect big starbursts to be happening, even among little galaxies ramming into each other. Once again, this is consistent with what we would expect.

Clearly, this Ultra Deep Field image builds upon the earlier Hubble Deep Field view in showing that there was a rapid evolution in the character of galaxies through starbursts and mergers during the first few billion years after the Big Bang. This observed evolution is completely consistent with the Big Bang model. But what about the very first stars and the very galaxies; if we can push Hubble any further, can we see those with it? Unfortunately, the very first stars, the very first galaxies in the universe are probably too faint for Hubble. One of the key reasons why is most of their light is redshifted into the infrared part of the spectrum where Hubble is not sensitive. One of the things that the Hubble Ultra Deep Field shows us is that even Hubble has its limits; even with spending 10 percent of all the Hubble time in a given year, we can't see all the first stars and the first galaxies in the universe.

In summary, the fundamental importance of the Hubble Ultra Deep Field is that it clearly shows that the universe of galaxies evolves over time. If the Ultra Deep Field had revealed that the most distant galaxies consisted of spirals and ellipticals just like those nearby, it would have been necessary to rethink galaxy evolution in the context of the Big Bang model. Instead, the Ultra Deep Field joins other independent pieces of evidence such as the microwave background radiation and Hubble's observations of the expanding universe in supporting the Big Bang as our best explanation for the origin and character of the universe. Next time, we'll conclude our lecture series with a discussion of Hubble's legacy and the challenges awaiting the next generation of large space telescopes in detecting extrasolar planets and the first galaxies. Please join us then.

Hubble's Legacy and Beyond
Lecture 12

> The Kepler Space Telescope will be able to detect Earth-size orbits, those orbits that have a period of a year or more. ... The Kepler telescope will answer a question that humanity has asked for thousands of years: Are Earth-sized planets rare or common around other stars in the galaxy?

Hubble is not capable of directly imaging an Earth-size or Jupiter-size planet in an Earth-size or Jupiter-size orbit around any solar-type star at optical wavelengths. Nevertheless, we are living in the golden age of **exoplanet** discovery right now. Since the early 1990s, more than 400 exoplanets have been discovered indirectly through spectroscopy on ground-based telescopes. The method for finding exoplanets uses our old friend the Doppler effect. You point a large telescope with a very efficient, high-resolution spectrograph at a star and you study it for very tiny velocity shifts. These shifts may be due to an orbiting planet tugging on the star, occasionally pulling it toward us and then away from us. The indirect gravitational effect of wobbling will be picked up as tiny blue and red shifts corresponding to the period of the planet's orbit.

The Doppler method has been amazingly successful: Planets have been found around about 15 percent of the 2,000-some solar-type stars surveyed. A little more than 20 years ago, the only planets we knew about in the entire galaxy were in our solar system. Another indirect approach to detecting planets is to study stars for planets with edge-on orbits: As the planet passes in front of the star, it dims the light of the star. This method has already had a great deal of success from the ground: More than 40 Jupiter-sized planets have been detected in close orbits through this method.

Although Hubble has not imaged an Earth or a Jupiter in comparable orbits around another star, it has optically detected an exoplanet in observations released in 2008. The targeted star was Fomalhaut, somewhat younger and hotter than the Sun and 25 light-years away. By comparing the 2004 and 2006 data, Hubble astronomers found a small dot, a point of light, just inside the dust belt around Fomalhaut. This dot, a billion times fainter than the star, is an exoplanet. Only because this exoplanet is three times more massive than Jupiter and very far away from its host star could Hubble find it. The direct detection of the Fomalhaut exoplanet is an impressive example of the limits of Hubble and the need for new technology to image less massive exoplanets much closer to their host stars.

The next big step in imaging extrasolar planets will take place in 2014 when NASA launches the successor to Hubble, the James Webb Space Telescope. The Webb telescope will have a mirror much bigger than that of Hubble (6.5 meters as opposed to 2.4), its optics and instruments will be optimized for observations at infrared wavelengths, and (to best study this infrared-wavelength light) its orbit will be outside the Earth's orbit. What that means is if something goes wrong with Webb, there's no way to fix it.

Until the Webb telescope is launched and, perhaps, afterward, Hubble will continue to explore the cosmos. The space shuttle servicing mission to

There will be no further upgrades or repairs of Hubble. What is the eventual fate of Hubble? Sometime beyond 2020, its orbit will decay, and if nothing is done, it will burn up in the atmosphere—at least most of it will, but some pieces could make it to the ground.

Hubble in 2009 was a fantastic success. New instrumentation was installed; the Space Telescope Imaging Spectrograph, which failed several years ago, was repaired and is working well now as a result of this servicing mission; and the **Advanced Camera for Surveys** was also restored almost to its full functionality. In addition to working on these instruments, the astronauts on the servicing mission performed important basic maintenance on the telescope. The scientific lifetime of Hubble has now been extended for at least five years and maybe even a decade.

However, the bad news is that there will be no further upgrades or repairs of Hubble. By retiring the space shuttle fleet, NASA will no longer be able to fix Hubble. Sometime beyond 2020, its orbit will decay and it will burn up in the atmosphere—at least most of it will, but some pieces could make it to the ground. Concerns about that led the astronauts of the 2009 servicing mission to attach a docking port to Hubble, so that at some point in the future, before its orbit decays, a robotic spacecraft might be sent up, attach itself to Hubble, and de-orbit Hubble into the ocean safely.

By retiring the space shuttle fleet, NASA will no longer be able to fix Hubble. Sometime beyond 2020, its orbit will decay and it will burn up in the atmosphere.

Until its eventual demise, all of Hubble's observations will continue to be saved in its vast public data archive at http://archive.stsci.edu. This archive has been a tremendous resource for researchers; indeed, many research papers have been based on the archive data alone. But beyond the raw data, Hubble's legacy will be the spectacular images that have not only helped researchers to rewrite the textbooks but have also stimulated students of all ages to learn more about the universe. ■

Important Terms

Advanced Camera for Surveys (ACS): The third-generation camera installed onboard HST during the 2002 space shuttle servicing mission.

exoplanet: A planet outside the solar system.

Suggested Reading

Boss, *The Crowded Universe.*

Gardner, "Finding the First Galaxies."

Mayor and Frei, *New Worlds in the Cosmos.*

Questions to Consider

1. What would be the implications for exoplanet studies and future NASA missions if the Kepler Space Telescope finds that Earth-sized exoplanets are rare?

2. Which HST images will have the most long-lasting scientific and aesthetic impact?

Hubble's Legacy and Beyond
Lecture 12—Transcript

Welcome back to our voyage of discovery at the cosmic frontier with the Hubble Space Telescope. Last time, we explored the Hubble Ultra Deep Field and its 13 billion year view of galaxy evolution packed into one cosmic image. Throughout this course, we have taken a long journey in time and space with Hubble, from a recent comet impact on the planet Jupiter to young galaxies on the doorstep of the Big Bang. Yet, the ten spectacular images of planets, stars, nebulae, and galaxies that we've focused on are just a tiny sample of the rich harvest of astronomical discovery provided by Hubble. In addition to resolving out the cocoons of star formation in the Eagle Nebula and probing the nature of dark energy through observations of distant supernovae, Hubble has also convincingly identified supermassive black holes at the centers of many galaxies, detected the atmospheres of exoplanets around other stars, and imaged seasonal variations on the surface of Pluto. Hubble has accomplished all of its work since 1990 while observing just a tiny fraction of the sky. Although there is still a lot in the cosmos for Hubble to explore, many of its discoveries have already stimulated new research questions that will require new space telescopes and technologies to successfully answer.

Cosmology and the study of exoplanets are currently the most active research areas on the astrophysics frontier and they are pushing the limits of Hubble's capabilities. For example, the success of the Ultra Deep Field also illustrates Hubble's limitations in imaging ever more distant galaxies. Specifically, the first galaxies and stars in the universe are most likely beyond Hubble's reach due to their faintness and the redshift of most of their radiation out of the optical into the infrared. In the case of exoplanets, much of this work has been pioneered through ground-based spectroscopic studies that have established that planets are indeed common around other stars. Although Hubble has had some success in studying the atmospheres of exoplanets spectroscopically, it faces real challenges in directly imaging these objects. In this final lecture, we discuss the future of Hubble and the next generation of space telescopes in the context of the detection and characterization of exoplanets.

In the case of directly imaging planets like Earth and Jupiter around other Sun-like stars, the problem has been described as actually much more difficult than detecting a firefly sitting on the edge of a searchlight. Why is this so hard? First, you need to realize, in the optical wavelengths, planets do not shine on their own; the only reason we see them with the naked eye is because the reflecting sunlight. Seen from afar, planets are very, very faint. Furthermore, when seen from afar, planets are very, very close to their home star. Let's think about this in the context of imagine observing the solar system from Alpha Centauri, the nearest star, 4.3 light-years away. As observed from Alpha Centauri, Earth would be two billion times fainter than the Sun, and it would be very close to the Sun: The separation in terms of angle between the Earth and the Sun as viewed from Alpha Centauri would be just three quarters of an arc second. You may think: Maybe it would be somewhat easier with Jupiter; it's a much bigger planet. But Jupiter is further away from the Sun, too; it turns out that at optical wavelengths, Jupiter is 300 million times fainter than Sun, and it's somewhat further away than Earth is as viewed from Alpha Centauri, but it would still be separated by only four arc seconds.

We can turn the problem around: Let's say we're here on Earth and imagine we're looking for an Earth-like planet in an Earth-type orbit around Alpha Centauri and a Jupiter-sized planet in a Jupiter-type orbit around Alpha Centauri. From the solar system looking at Alpha Centauri, the numbers I just gave you would be the same for Alpha Centauri; and the problem gets even harder towards more distant stars. Indeed, if Earth is separated by three-quarters of an arc second around Alpha Centauri, an Earth around a more distant star would be an even tinier angle; as the distance goes up, the angular separation gets smaller and smaller and smaller on our sky. To put all this into perspective, recall the image I showed you earlier in the course, the Hubble image of the brightest star in the sky, Sirius. If you recall, Sirius has a faint white dwarf companion; the white dwarf companion of Sirius is about 10,000 times fainter than Sirius itself. And it's close; only seven arc seconds separate the white dwarf companion of Sirius to Sirius. Look at this Hubble image and stare at this, and then think about what it would be like to get an Earth imaged around Alpha Centauri. In that case, the Earth around Alpha Centauri would be 10 times closer than this separation of the white dwarf from Sirius, and Earth would be even 200,000 times fainter than the white dwarf is in this particular image of Sirius. The bottom line is that Hubble is not capable of directly imaging an Earth-size or Jupiter-size planet in an Earth-size or Jupiter-size orbit around any solar-type star at optical wavelengths. The imaging problem would be somewhat less difficult working at infrared wavelengths because the brightness difference at infrared wavelengths between a planet and orbiting star somewhat less than it is at optical wavelengths; and, of course, it would also be easier if the planets were orbiting in a much larger orbit than, say, Jupiter. But nevertheless, the bottom line is Hubble Space Telescope simply cannot detect Earths or Jupiters around other stars.

Now you might say: Well wait a minute; if imaging planets with our best telescope, the Hubble Space Telescope, is so challenging, what is the evidence of solar-system-type extrasolar planets? Actually, prior to the early 1990s, the only planets known in the entire universe were those in the solar system. This is worth just thinking about for a moment. The naked-eye planets we can see in a night sky—Mercury, Venus, Mars, Jupiter, and Saturn—have been known since ancient times thousands of years ago. The only planets that have been discovered by humanity since these ancient times thousands of years ago and the early 1990s were Uranus, Neptune, and Pluto (which was then subsequently demoted to a minor planet; we'll move that issue aside). But the point is, since the early 1990s, over 400 exoplanets have been discovered. We are living in the golden age of exoplanet discovery right now. The vast majority of these exoplanets have been discovered indirectly thru spectroscopy on ground-based telescopes. I want to spend a minute or two describing this method how the vast majority of these planets have been found.

Once again, it utilizes our old friend the Doppler Effect. You point a large telescope with a very efficient, high-resolution spectrograph at a star and you study its motion; you look at absorption lines in its spectra and you see if it's moving. You look at it with very high position so you can see very tiny velocity shifts; and you may see these velocity shifts, and what they can be due to is a planet orbiting the star. As the planet orbits a star, the planet will tug on the star. As the planet tugs on the star, the star will occasionally be moving toward us and then away from us; we'll see tiny blue shifts as it moves toward us and tiny red shifts as it moves away, tiny little velocity shifts that will correspond to the period of that planet orbiting around the star. I want to emphasize: We do not see the planet; we just see its indirect effects, its indirect gravitational effects, on the star wobbling. Note that this is a velocity effect measured with spectroscopy: The effect in terms of its position on the sky changing is so tiny we could never measure it, certainly not even with Hubble. Our spectrographs are sensitive enough now that we can detect velocity changes of a planet tugging on its host star equivalent to a brisk walk of a human being. Given this sensitivity, what kind of planets can we detect with this Doppler method of detection? It turns out they're most sensitive to massive planets like Jupiter and Saturn, or more massive than those; because the more massive planet will cause a bigger tug on the star and we'll see a bigger velocity shift that is easier to measure with our spectrographs. Particularly, the effect is most notable in those massive planets that have short periods; those will really stand out in this Doppler method.

Because the Doppler method is so sensitive to the most massive planets, that partly explains at least why over 80 percent of the planets so far through the Doppler method have masses greater than that of Saturn; the method is simply most sensitive to massive planets. Thus it can't say much about lower-mass planets, certainly Earth-mass planets; indeed, the Doppler method has yet to detect an Earth-mass planet around a solar-type star simply because an Earth-mass planet doesn't give enough of a tug yet so that we can see the very tiny velocity shift. The method, number one, is not that sensitive to planets certainly Earth-mass and below Earth-mass, and even planets that are a few Earth-masses in mass. Also this method is not that sensitive yet to planets that have large orbits and correspondingly long periods to go around their star; simply because these observations have only been going on since the early 1990s and the longer-period planets would have orbits on the order of decades, we simply haven't been observing these stars long enough to actually see these longer-period planets. The good news is the Doppler Method on our best and biggest ground-based telescopes and with ever more sensitive spectrographs is nearing the sensitivity to detect Earths around solar-like stars. But despite the limitations the Doppler Method has, it's been amazingly successful: To date, planets have been found around a little over 15 percent of the 2,000-some solar-type stars surveyed; and given the planets that are missing, the types of planets that wouldn't be found by this method so far, the true percentage of stars that may have planets may be actually higher than 50 percent. The bottom line is the Doppler Method has given astronomers tremendous encouragement that more than half the stars in the galaxy have planets around them; and literally, think about it, no more than a little over 20 years ago, the only planets we knew about in the entire galaxy were in our solar system. It just shows how fast science can change as technology develops.

There's another indirect approach to detecting planets besides the Doppler approach, and this approach is really neat; it actually follows up the idea of solar eclipses in a sense. The idea is you study a star—actually you study a large group of stars—on the sky; and if they have planets around them, and if those planets are orbiting in our line of sight, when that planet passes in front of the star we'll see a small dip in the light of the star because that planet will simply eclipse part of the star. You might say: What fraction of stars will have planets in exactly these edge-on orbits? You can estimate statistically this: The way you want to do this method is you don't want to study a large sample of stars, and consider that the orbits of a planet around a star will have a whole range of inclinations. Of course, if you have a star like this and a planet's going around it you won't see an eclipse, you'll only see it when it goes like this; but if you look at a large enough sample of stars, some significant fraction will have these edge-on orbits, and you can think about looking for these eclipses, these transits of the planet, as they pass in front of the star.

How much of a dimming will you get if this occurs? In the case of a solar-type, solar-sized host star, if it has a Jupiter passing in front of the star that would be enough to dim the light of that star by about one percent. An Earth-sized transit is much less noticeable: It would dim the light only .01 percent (by one part in 10,000). The transit method in terms of sensitivity would also be most sensitive to detecting the transits of Jupiter-sized planets rather than Earth-sized planets around stars. This method has already had a great deal of success from the ground: Over 40 Jupiter-sized planets have actually been detected so far in close orbits through this method, and the method, though, from the ground is not sensitive enough yet for Earth exoplanets. Why? It's because the sensitivity you need: You need a sensitivity of $1/10,000^{th}$ dips; to detect those little tiny dips in the light with our atmosphere—we've talked about the problems associated with our atmosphere before—but trying to make very accurate measurements of the light from a star to 1 part in 10,000 from ground-based observations is really hard to do. There's another key point, and that is you need observation continuity, you need to keep observing, if you want to get a longer orbit for a planet due to a transit. The point is: If you have a planet orbiting a star with a period of about one year, the transit will only occur once every year; so in order to make sure you catch it the first time, you just have to keep observing it continuously, and that's hard to do on the Earth because clouds and you'd need to have observatories all around the Earth and it's really complicated to do.

What you basically need is a dedicated transit space observatory; and to do this, NASA launched the Kepler space telescope in 2009. The Kepler telescope was a great idea. What it is, it's a space telescope with a primary mirror with a diameter of 1.4 meters, and its mission is to continuously monitor 100,000 stars in the constellation Cygnus. It's just going to stare at these same 100,000 stars for four years; just taking snapshot after snapshot after snapshot after snapshot with its electronic camera continuously. Its electronic cameras will have the sensitivity to detect the tiny little brightness dip of an Earth transit. Furthermore, with its four-year mission, it will be able to detect Earth-size orbits; those orbits that have a period of a year or more. What it means is if Earths are common, Kepler will detect hundreds; if Earths are not common, Kepler might not find any. The bottom line is that the Kepler telescope will answer a question that humanity has asked for thousands of years: Are Earth-sized planets rare or common around other stars in the galaxy? When the Kepler mission ends, we will know the answer to this fundamental question. That's one of the amazing things about this exoplanet work and how fast it's changed over the past 20 years. These are questions that humanity's wondered about for thousands of years, and we're on the threshold of getting the answers to these kinds of questions.

The success of the Doppler and transit methods of indirectly detecting planets around other stars continues to drive efforts to image them directly. It's one thing to see them indirectly, to see evidence of them, and we can learn a lot from these indirect methods; but we just want to see them, we'd like to be able to image them. We've already talked about that although Hubble is not capable of imaging an Earth or a Jupiter in comparable orbits around another star, it actually has provided the first optical detection of an exoplanet in observations released in 2008. The targeted star was Fomalhaut. Fomalhaut is somewhat younger and hotter than the Sun; it's a distance of about 25 light-years. The image you see here is a composite of data taken from observations taken in 2004 and 2006. Its total exposure time in this image is about 11 hours. It was taken with the ACS camera in coronagraphic mode. What does that mean? This instrument with the ACS, what it does is it tries to block the light from the central star so you can look for anything illuminated around the star. The instrument has its limits because it doesn't completely block the central starlight. Some of that light leaks through a little bit and is scattered into the image; you can see that a lot of the noise in this image is scattered starlight.

But despite that noise, you can see real features in this image. Specifically, you see kind of a large elliptical ring of light around this dark spot in the middle. This large elliptical ring of light is actually scattered light off a large dust belt around Fomalhaut. It was actually first detected through infrared observations in the 1980s due to the heat associated with these little dust grains. What the Hubble astronomers also noted that was fascinated is in studying this image very closely and comparing the 2004 and 2006 data, they found a little dot, a little point of light, just inside the dust belt. This little dot is a billion times fainter than Fomalhaut. It is an exoplanet; they've actually imaged an exoplanet. It's this little dot here, a billion times fainter than Fomalhaut. The distance this little dot is from the star corresponds to 20 times farther away than Jupiter is from the Sun. The mass associated with this exoplanet is estimated to be three times that of Jupiter. The data indicates the orbital period of the planet is 870 years; that's over 3 times that of Pluto around the Sun. The bottom line is this exoplanet is more massive than Jupiter, 3 times more massive, and it's very, very, very far away from its host star; that's why Hubble had the ability to pull it out of the murk here and find this little dot representing this exoplanets. The direct detection of the Fomalhaut exoplanet is an impressive example—it's an impressive result, but it's also an impressive example—of the limits of Hubble and the need for new technology to image less massive exoplanets much closer to their host stars.

The next big step in imaging extrasolar planets will take place in 2014 when NASA launches the successor to Hubble, the James Webb Space Telescope. The Webb Telescope will have a mirror much bigger than that of Hubble: Its mirror will be 6.5 meters in diameter, which will have seven times the light-gathering power than the 2.4 meter diameter mirror of Hubble. It's also a different kind of design. Hubble's mirror is basically a big hunk of glass; but the Webb Telescope will have a segmented mirror that after its launch it will actually open up to the appropriate figure to collect starlight. The optics and instruments on board the Webb Telescope also will be different from Hubble in the sense that they will be optimized for observations at infrared wavelengths; and this is very important because by having the instrumentation optimized for infrared, the Webb Telescope will be able to do science and go beyond the limits that we currently have with Hubble into new areas, specifically we'll be able to image exoplanets with the Webb Telescope better than we can with Hubble because exoplanets will typically emit more light in infrared wavelengths than at optical wavelengths. Also as we talked about before, the best hope to detect the first stars and the first galaxies in the universe is to observe them at infrared wavelengths, and that's also within the possibility of the Webb Telescope's capabilities.

Since it's working at infrared wavelengths, the Webb Telescope is equipped with a large sunshield, and it needs this to block the infrared radiation from the Sun, the Earth, and the Moon. If you're going to work in the infrared, you have to worry about stray heat radiation; and so this sun shield will block that light. In order to optimize your capability of being sensitive to infrared wavelength light, you want to try to get as far away as you can from Earth and the Moon; so that's why the position of the Webb Telescope will be quite different than that from Hubble: The Webb Telescope will be positioned 1.5 million kilometers away from Earth, outside the Earth's orbit; and from that position, as it orbits the Sun, the Sun, the Moon, and the Earth will be in the same direction so the sun shield of the Webb Telescope can block the infrared light. Also note that the Webb Telescope will be much further away from Earth than Hubble; what that means is if there was a space shuttle, to service it, it wouldn't be able to go there. One of the advantages of Hubble, of course, in the lower-Earth orbit, is that the space shuttle can constantly go up and fix when things go wrong and replace the instrumentation. But even if NASA was not retiring the space shuttle fleet, the Webb Telescope would be located far too far away to send any servicing mission to it. What that means is if something goes wrong with Webb, that's it; there's no way to fix it. That's kind of a drawback to this particular mission; but on the other hand, to do the kind of infrared work that Webb is designed to do, you really need to put it where it's going to be, 1.5 million kilometers from the Earth.

Once of the types of science that Webb will be capable of doing is it will have its own infrared coronagraph in terms of instrumentation. It will be able to take images around stars, and with this coronagraph it will block that starlight and look in the infrared to look for planets around the star and maybe dust rings as well. In the case of the Webb Telescope, with this instrumentation it could easily image the planet around Fomalhaut and its dust ring. Indeed, the Webb Telescope will be much more sensitive than the current infrared space telescope that NASA has up in orbit, and that's the Spitzer 0.85 meter telescope that has already revealed lots of wonderful things in the infrared about stars and galaxies; the Webb Telescope will do much more than Spitzer, and in the context of exoplanets it should actually be able to detect Jupiters much closer-in to their nearby stars, although probably not as close as Jupiter is to our Sun.

The grand dream of exoplanet studies, of course, is to image and characterize Earth-type exoplanets. This could possibly be realized as early as the late 2020s through the launch of a space telescope called the Terrestrial Planet Finder Interferometer, or the TPF-I. This is an amazing idea. TPF-I will consist of several 3-meter diameter mirrors, and they will fly in formation at some position far from Earth and they will be spread in space over a baseline of on the order of 100 meters. What this group of telescope mirrors will do: When they point at the same object, the light from each one of these mirrors will be focused into one image. What's neat about an interferometer like this is that together such separate mirrors can act as one big telescope that's the size of the baseline separation of the individual mirrors. What that means is these mirrors will effectively act as a 100-meter infrared space telescope. It will end up having a much higher resolving power than the Webb Telescope. Specifically, the TPF-I could image an Earth around a nearby star.

When I say "image," what I mean is the image, of course, would still only be a dot. You might say, "A dot, what can we do with a dot? I want to image a planet, I want to characterize; what is a dot going to do for me?" Remember, we can learn a lot from dots; after all even with Hubble almost all the stars in the sky are dots. You can take spectra of dots, you can see if dots vary in brightness, and we've learned tremendous amounts about stars simply as studying them as dots. So even if all we image of a nearby exo-Earth is a dot, we could learn some enormous things about it. Specifically, TPF-I could take an infrared spectrum of this dot and compare it to the infrared spectra of the planets we know about in the solar system—the Earth, Venus, and Mars— and compare: What does this exo-Earth spectrum look like as compared to the terrestrial kinds of planets we know about in our solar system? Such a spectrum would allow us to study the atmosphere of this exo-Earth. It could reveal gaseous compounds in the atmosphere of this planet such as water, ozone, methane, carbon dioxide, all kinds of things. The importance of this is that the relative mix of these gases could indicate the presence of life on that planet. For example, the presence of oxygen in an atmosphere is kind of unusual; oxygen reacts very quickly with most things and there shouldn't be a stable amount of oxygen in an atmosphere unless you have life. There could be other explanations, but if you have oxygen in an atmosphere of a planet, that's one key indicator that life may be there. The bottom line is although the vast distances to even the nearest exo-Earths will preclude us from physically travelling to these worlds for the far foreseeable future, their observations with a space telescope such as the TPF-I might well give us with our best first evidence of extraterrestrial life beyond the solar system.

Until the Webb Telescope is launched, and perhaps beyond, Hubble will continue to explore the cosmos. The space shuttle servicing mission to Hubble in 2009 was a fantastic success. New instrumentation was installed including the Widefield Camera Three (WFC3) that will enhance Hubble's near-infrared sensitivity; its ability to detect objects at near-infrared wavelengths. Also another new instrument, the Cosmic Origin Spectrograph, was installed, and this spectrograph will be capable of taking images at ultraviolet wavelengths from very faint sources far away in the distant universe and use those to study gas clouds at great distances between us and these ultraviolet sources and probe more not only the large-scale structure of galaxies in the universe but the large-scale structure of gas clouds in the universe. In addition, the instrument that I use on board Hubble, STIS, the Space Telescope Imaging Spectrograph, which failed several years ago, was repaired and is working great now as a result of this servicing mission; and the ACS camera was also restored almost to its full functionality. In addition to working on these instruments, the astronauts during the servicing mission also did a lot of very important basic maintenance of the telescope: They replaced the gyroscopes; they replaced the batteries. Hubble's batteries hadn't been replaced since it was launched, and just like your cell phone battery degrades after charge and recharge and recharge and recharge, the batteries were beginning to go on Hubble. Indeed, if that servicing mission in 2009 had not been launched, Hubble's scientific lifetime probably would have been over in another year or two; but as a result of this mission, the scientific lifetime of Hubble has now been extended for at least five years and maybe even a decade.

However, the bad news is there will be no further upgrades or repairs of Hubble. By retiring the space shuttle fleet, there's no longer any way for NASA to get up there and fix Hubble. Next time something breaks on Hubble, it's over; we won't be taking any more scientific observations. But the hope is there's a strong feeling it should definitely survive until the launch of the Webb Telescope, and hopefully beyond it. But what is the eventual fate of Hubble? Sometime beyond 2020, its orbit will decay, and if nothing is done it will burn up in the atmosphere; at least most of it will, but some pieces could make it to the ground. Concerns about that led to one of the things the astronauts did during the 2009 service mission: attaching a docking port to Hubble in its back, such that at some point in the future before its orbit decays, a robotic spacecraft might be sent up, attach itself to Hubble, and de-orbit Hubble into the ocean safely. This isn't satisfying all, of course; Hubble belongs in the Smithsonian Air and Space Museum after its life is done, but that just isn't going to happen.

Until its eventual demise, all of Hubble's observations will continue to be saved in its vast public data archive at http://archive.stsci.edu. If we look at an image showing points of light where Hubble's already observed objects in the sky, we see there's a lot of open space in this image; in other words, there's still a lot of sky for Hubble to observe. This archive has been a tremendous resource for researchers; many research papers indeed based on the archive data alone. Anyone can use this archive to explore the cosmos; it's all open to the public. As soon as astronomers typically get a year to work on their data individually, it goes right into this archive; so there's a vast amount of data in this archive, some of it that really hasn't been studied much at all.

But beyond the raw data, Hubble's legacy will be the spectacular images that have not only helped researchers to rewrite the textbooks but have also stimulated students of all ages to learn more about the universe. Just as a work of art can speak deeply to us, these images convey a sense of wonder and the spirit of discovery. Whatever images you found most appealing in this course, I hope that you have enjoyed our brief journey through Hubble's universe. Undoubtedly, there will be new candidates for Hubble's top 10 images as it continues to unveil the cosmic frontier.

Glossary

Advanced Camera for Surveys (ACS): The third-generation camera installed onboard HST during the 2002 space shuttle servicing mission.

asteroid: Member of a class of rocky objects orbiting the Sun, ranging in size from meters to hundreds of kilometers.

baryon: Member of a class of particles, including protons and neutrons, that make up most of the matter in the visible universe.

Big Bang universe: An evolving universe that had a singular origin in time.

black hole: A region of severely curved space around a collapsed stellar core where not even light can escape.

Cassegrain telescope: A telescope in which incoming starlight is reflected off a primary mirror to a secondary mirror that then reflects it back through a small central hole in the primary mirror to an eyepiece or instrument behind the primary.

Cepheid variable: A type of pulsating star with a period-luminosity relation that is useful in determining distances to the star's host galaxy.

comet: A kilometer-sized object of ice and rock that produces a visible tail of vapor and dust as it approaches the Sun during the course of its orbit.

cosmic microwave background (CMB) radiation: The redshifted radiation from a very early time (380,000 years after the Big Bang) when the expanding universe transitioned from its hot, dense, bright origin to a cooler, transparent state.

dark energy: The mysterious energy driving the observed acceleration in the expansion of the universe.

dark matter: The dominant, unknown constituent of matter in the universe that interacts with visible matter gravitationally but not through electromagnetic forces.

Doppler effect: The wavelength shift in the spectrum of a light source as that source moves toward or away from an observer.

Einstein ring: The simplest case of a gravitational lens, in which the observer, lens, and background object are perfectly aligned.

electromagnetic radiation: Commonly referred to as "light" at optical wavelengths, this radiation is due to oscillating electric and magnetic fields.

elliptical galaxy: A gas-poor, elliptically shaped system of up to 500 billion typically old stars.

emission nebula: A glowing cloud of interstellar gas heated by the ultraviolet light of a nearby hot star.

exoplanet: A planet outside the solar system.

globular star cluster: A densely packed, spherical cluster of up to a million old stars typically found in the halos of galaxies.

gravitational lens: An object (or cluster of objects) whose mass is large enough to curve the surrounding space to a degree at which distortions are produced in the images of background objects.

Hertzsprung-Russell (HR) diagram: A diagram comparing the temperatures and luminosities of stars that is useful in charting their evolution over time.

Hubble's constant: The constant of proportionality between the redshift velocities of galaxies and their distances.

Hubble's law: The linear relationship between the redshift velocities of galaxies and their distances that is indicative of an expanding universe.

interstellar dust: Submicron-sized interstellar particles of carbon, oxygen, and silicon compounds that are effective in attenuating background starlight.

interstellar molecular cloud: An interstellar cloud of gas and dust that is dense enough to form molecules (and sometimes stars) and block the optical light of background stars.

light-year: The distance (6 trillion miles) that light travels in one year.

main sequence star: A star that is powered by the nuclear fusion of hydrogen into helium in its core.

Near-Earth Object (NEO): A nearby asteroid whose orbit intersects that of Earth.

neutrino: An elementary particle of very low mass produced in nuclear reactions, such as those in the solar core and Type II supernovae.

neutron star: The collapsed core remnant of a Type II supernova; it typically has a mass of about 1.5 solar masses within its radius of 10 kilometers.

photon: The particle that carries electromagnetic radiation (light) with wavelike characteristics.

planetary nebula: The short-lived nebula (lasting for about 50,000 years) that results when a dying red giant blows off the gaseous layers surrounding its core.

pulsar: A rapidly rotating neutron star.

red giant: After a solar-type main-sequence star fuses all of its core hydrogen into helium, it evolves into this type of cooler, larger, more luminous star.

resolving power: A measure of the smallest angular separation that a telescope can resolve in an image.

seeing: A measure of the limiting resolving power for ground-based telescopic observations produced by atmospheric turbulence.

spherical aberration: An error in the shape of a lens or mirror that prevents all of the incident light from coming into focus at the same imaging point.

spiral galaxy: A gas-rich, disk-shaped system of young and old stars with ongoing star formation in its characteristic spiral arms.

standard candle: A class of objects whose luminosity is assumed to be known.

Steady State universe: A non-evolving universe whose large-scale characteristics are constant in space and time.

supergiant: After a massive main-sequence star fuses all of its core hydrogen into helium, it eventually evolves into this type of very large, very luminous, cool star.

supernova remnant: The expanding, nucleosynthetically enriched gaseous remnant of a star that has undergone a supernova explosion.

white dwarf: The final stage in the evolution of the Sun (and 99 percent of the stars in the Milky Way Galaxy); a compact, Earth-sized object radiating its remnant energy like a dying coal in a fireplace.

Wide-Field Camera 3 (WFC3): The fourth-generation camera installed onboard HST during the 2009 space shuttle servicing mission.

Wide-Field Planetary Camera 2 (WFPC2): The second-generation camera installed onboard HST during the 1993 space shuttle servicing mission.

Bibliography

General Background Materials:

Bartusiak, Marcia. *Archives of the Universe: 100 Discoveries That Transformed Our Understanding of the Cosmos.* New York: Vintage Books, 2004. A masterful compilation of the original discovery papers of astrophysical phenomena ranging from exoplanets to the cosmic microwave background radiation—all presented with insightful introductions for the general reader.

Devorkin, David, and Robert W. Smith. *Hubble: Imaging Space and Time.* New York: Random House, 2008. A beautiful compilation and discussion of the most spectacular images taken by the HST.

Hester, Jeff, David Burstein, George Blumenthal, Ronald Greeley, Bradford Smith, and Howard G. Voss. *21st Century Astronomy*, 2nd ed. New York: W.W. Norton, 2007. A thorough, well-illustrated introductory textbook on astronomy.

Pasachoff, Jay M., and Alex Filippenko. *The Cosmos: Astronomy in the New Millennium*, 3rd ed. Belmont, CA: Thomson Brooks/Cole, 2007. A thorough, well-illustrated introductory textbook on astronomy.

Lecture-Specific Materials:

Balick, Bruce. "How the Sun Will Die." *Astronomy* 36 (December 2008): 38–43. An informative, well-illustrated popular article about the evolution of solar-type stars and the wide variety of planetary nebulae observed in the Milky Way Galaxy.

Bartusiak, Marcia. *The Day We Found the Universe*. New York: Pantheon Books, 2009. A wonderful, well-researched history of the scientific efforts of Edwin Hubble, Henrietta Leavitt, Vesto Slipher, and others that led to the recognition of spiral nebulae as distant galaxies and the discovery of the expanding universe.

Boss, Alan. *The Crowded Universe: The Search for Living Planets*. New York: Perseus Books, 2009. An insider's review of the science and history of recent efforts to discover and characterize exoplanets.

Christensen, Lars Lindberg, Davide de Martin, and Raquel Yumi Shida. *Cosmic Collisions: The Hubble Atlas of Merging Galaxies*. New York: Springer, 2009. A well-illustrated atlas and overview of the colliding galaxies observed with the HST.

Christianson, Gale E. *Edwin Hubble: Mariner of the Nebulae*. New York: Farrar, Straus, and Giroux, 1995. A detailed biography of the life and science of Edwin Hubble.

Crovisier, Jacques, and Therese Encrenaz. *Comet Science: The Study of Remnants from the Birth of the Solar System*. Cambridge, UK: Cambridge University Press, 2000. A concise introduction to the scientific study of comets.

Freeman, Ken, and Geoff McNamara. *In Search of Dark Matter*. New York: Springer, 2006. A concise introduction to the astrophysics of dark matter.

Gardner, Jonathan P. "Finding the First Galaxies." *Sky and Telescope* 119 (January 2010): 24–30. An informative, well-illustrated popular article about the Hubble Ultra Deep Field and future efforts to study the first galaxies with the James Webb Space Telescope.

Gates, Evalyn. *Einstein's Telescope: The Hunt for Dark Matter and Dark Energy in the Universe*. New York: W.W. Norton, 2009. A well-written introduction to the astrophysics of gravitational lenses, dark matter, and dark energy.

Kaler, James B. *Cosmic Clouds: Birth, Death, and Recycling in the Galaxy*. New York: W.H. Freeman, 1997. A well-written, well-illustrated introduction to the cycling of interstellar gas and dust through star formation and stellar evolution in the Milky Way Galaxy.

————. *Extreme Stars: At the Edge of Creation*. Cambridge, UK: Cambridge University Press, 2001. A well-written introduction to the astrophysics of stars and the Hertzsprung-Russell diagram.

————. *The Hundred Greatest Stars*. New York: Springer, 2002. A well-illustrated introduction to the wide variety of stars in the Milky Way Galaxy.

Kirshner, Robert P. *The Extravagant Universe: Exploding Stars, Dark Energy, and the Accelerating Cosmos*. Princeton, NJ: Princeton University Press, 2004. A well-written insider's review of the science and history behind the discovery of dark energy.

Kwok, Sun. *Cosmic Butterflies: The Colorful Mysteries of Planetary Nebulae*. Cambridge, UK: Cambridge University Press, 2001. A well-illustrated introduction to the astrophysics of planetary nebulae.

Levy, David H. *Impact Jupiter: The Crash of Comet Shoemaker-Levy 9*. Cambridge, MA: Basic Books, 1995. The inside story of Comet Shoemaker-Levy 9 by a co-discoverer of the comet.

Loeb, Abraham, and T. J. Cox. "Our Galaxy's Collision with Andromeda." *Astronomy* 36 (June 2008): 28–33. An informative, well-illustrated popular article about the future collision between the Milky Way and Andromeda galaxies.

Mayor, Michel, and Pierre-Yves Frei. *New Worlds in the Cosmos: The Discovery of Exoplanets*. Cambridge, UK: Cambridge University Press, 2003. A well-written insider's introduction to the astrophysics of exoplanets by the co-discoverer of the first exoplanet around a solar-type star.

McCray, W. Patrick. *Giant Telescopes: Astronomical Ambition and the Promise of Technology*. Cambridge, MA: Harvard University Press, 2004. A well-written history of the science and politics behind the development and construction of large ground-based optical telescopes in the United States since the 1950s.

O'Dell, C. Robert. *The Orion Nebula: Where Stars Are Born*. Cambridge, MA: Harvard University Press, 2003. A well-written introduction to the study of star formation in the Orion Nebula by one of the founding scientists of the HST.

Singh, Simon. *Big Bang: The Origin of the Universe*. New York: HarperCollins, 2004. An excellent, well-written introduction to the history and astrophysics of Big Bang cosmology.

Smith, Robert W. *The Space Telescope*. New York: Cambridge University Press, 1989. A well-researched pre-launch history of the science and politics involved in the design, funding, and development of the HST.

Sparke, Linda S., and Jay S. Gallagher III. *Galaxies in the Universe*, 2nd ed. Cambridge, UK: Cambridge University Press, 2007. A detailed introductory textbook on the structure and astrophysical characteristics of the Milky Way and other galaxies.

Waller, William H., and Paul W. Hodge. *Galaxies and the Cosmic Frontier*. Cambridge, MA: Harvard University Press, 2003. A detailed introduction to the astrophysics of galaxies.

Wheeler, J. Craig. *Cosmic Catastrophes: Exploding Stars, Black Holes, and Mapping the Universe*, 2nd ed. Cambridge, MA: Cambridge University Press, 2007. A thorough introduction to the astrophysics of stellar evolution and supernovae.

Zimmerman, Robert. *The Universe in a Mirror: The Saga of the Hubble Space Telescope and the Visionaries Who Built It*. Princeton, NJ: Princeton University Press, 2008. An excellent behind-the-scenes history of the HST from its original idea through its many discoveries in orbit.

Advanced Background Materials:

AbdelSalam, Hanadi M., Prasenjit Saha, and Liliya L. R. Williams. "Nonparametric Reconstruction of Abell 2218 from Combined Weak and Strong Lensing." *The Astronomical Journal* 116 (October, 1998): 1541–1552. A detailed research article about the mass distribution of the galaxy cluster Abell 2218 as revealed by the HST image of its gravitational lensing.

Balick, Bruce, and Adam Frank. "Shapes and Shaping of Planetary Nebulae." *Annual Review of Astronomy and Astrophysics* 40 (2002): 439–486. A detailed research review of the observational and theoretical studies of the shapes of planetary nebulae.

———, Jeanine Wilson, and Arsen R. Hajian. "NGC 6543: The Rings around the Cat's Eye." *The Astronomical Journal* 121 (2001): 354–361. A detailed research article about the Cat's Eye Nebula as revealed by HST imaging.

Beckwith, Steven V. W., et al. "The Hubble Ultra Deep Field." *The Astronomical Journal* 132 (November 2006): 1729–1755. A detailed research article reporting the initial results and conclusions from the Hubble Ultra Deep Field image.

Hammel, H. B., et al. "HST Imaging of Atmospheric Phenomena Created by the Impact of Comet Shoemaker-Levy 9." *Science* 267 (March 3, 1995): 1288–1296. A detailed research article about the impact of Comet Shoemaker-Levy 9 on Jupiter's atmosphere as revealed by HST imaging.

Hester, J. Jeff. "The Crab Nebula: An Astrophysical Chimera." *Annual Review of Astronomy and Astrophysics* 46 (2008): 127–155. A detailed review of the latest observational research concerning the nature of the Crab Nebula.

———, et al. "Hubble Space Telescope WFPC2 Imaging of M16: Photoevaporation and Emerging Young Stellar Objects." *The Astronomical Journal* 111 (June 1996): 2349–2360. A detailed research article about the star formation in the inner Eagle Nebula as revealed by HST imaging.

Kalas, Paul, et al. "Optical Images of an Exosolar Planet 25 Light Years from Earth." *Science* 322 (November 13, 2008): 1345–1348. A detailed research article about the exoplanet discovered around the star Fomalhaut through HST imaging.

Kuijken, Konrad, and R. Michael Rich. "Hubble Space Telescope WFPC2 Proper Motions in Two Bulge Fields: Kinematics and Stellar Populations of the Galactic Bulge." *The Astronomical Journal* 124 (October 2002): 2054–2066. A detailed research article about the stellar population of the Galactic Bulge as revealed by the HST image of the Sagittarius Star Cloud.

Mengel, S., et al. "Young Star Clusters in Interacting Galaxies—NGC 1487 and NGC 4038/4039." *Astronomy and Astrophysics* 489 (October 2008): 1091–1105. A detailed research article about the young star clusters in the HST image of the Antennae Galaxies.

Riess, Adam, G., et al. "Cepheid Calibrations from the Hubble Space Telescope of the Luminosity of Two Recent Type Ia Supernovae and a Redetermination of the Hubble Constant." *The Astrophysical Journal* 627 (July 10, 2005): 579–607. A detailed research article about Cepheid variable stars, Type Ia supernovae, the Hubble constant, and the HST image of the galaxy NGC 3370.

Spitler, Lee R., et al. "Hubble Space Telescope ACS Wide-Field Photometry of the Sombrero Galaxy Globular Cluster System." *The Astronomical Journal* 132 (October 2006): 1593–1609. A detailed research article about the globular star clusters of the Sombrero Galaxy as revealed by HST imaging.

Notes

Notes

Notes

Notes

Notes